MAPPING FORESTRY

PETER EREDICS, editor

ESRI Press, 380 New York Street, Redlands, California 92373-8100

Compilation and design copyright © ESRI

All rights reserved. First edition 2010

14 13 12 11 10   2 3 4 5 6 7 8 9 10

Printed in the United States of America

Library of Congress Cataloging-in-Publication Data

Mapping forestry / Peter Eredics, editor.—1st ed.

    p. cm.

    ISBN 978-1-58948-209-8 (pbk. : alk. paper)

    1. Forest monitoring. 2. Forest management. 3. Geographic information systems. I. Eredics, Peter, 1970-

SD431.M27 2010

634.9'2—dc22   2009041761

The information contained in this document is the exclusive property of ESRI. This work is protected under United States copyright law and the copyright laws of the given countries of origin and applicable international laws, treaties, and/or conventions. No part of this work may be reproduced or transmitted in any form or by any means, electronic or mechanical, including photocopying or recording, or by any information storage or retrieval system, except as expressly permitted in writing by ESRI. All requests should be sent to Attention: Contracts and Legal Services Manager, ESRI, 380 New York Street, Redlands, California 92373-8100, USA.

The information contained in this document is subject to change without notice.

U.S. Government Restricted/Limited Rights: Any software, documentation, and/or data delivered hereunder is subject to the terms of the License Agreement. In no event shall the U.S. Government acquire greater than restricted/limited rights. At a minimum, use, duplication, or disclosure by the U.S. Government is subject to restrictions as set forth in FAR §52.227-14 Alternates I, II, and III (JUN 1987); FAR §52.227-19 (JUN 1987) and/or FAR §12.211/12.212 (Commercial Technical Data/Computer Software); and DFARS §252.227-7015 (NOV 1995) (Technical Data) and/or DFARS §227.7202 (Computer Software), as applicable. Contractor/Manufacturer is ESRI, 380 New York Street, Redlands, California 92373-8100, USA.

ESRI, ESRI Press logo, www.esri.com, @esri.com, ArcGIS, ArcMap, 3D Analyst, ArcScripts, ArcToolbox, ArcEditor, ArcCatalog, Maplex, ModelBuilder, ArcInfo, ArcPad, ArcReader, ArcView, ArcGlobe, and ArcSDE are trademarks, registered trademarks, or service marks of ESRI in the United States, the European Community, or certain other jurisdictions. Other companies and products mentioned herein are trademarks or registered trademarks of their respective trademark owners.

Ask for ESRI Press titles at your local bookstore or order by calling 800-447-9778, or shop online at www.esri.com/esripress. Outside the United States, contact your local ESRI distributor or shop online at www.eurospanbookstore.com/ESRI.

ESRI Press titles are distributed to the trade by the following:

In North America:
Ingram Publisher Services
Toll-free telephone: 800-648-3104
Toll-free fax: 800-838-1149
E-mail: customerservice@ingrampublisherservices.com

In the United Kingdom, Europe, Middle East and Africa, Asia, and Australia:
Eurospan Group
3 Henrietta Street
London WC2E 8LU
United Kingdom
Telephone: 44(0) 1767 604972
Fax: 44(0) 1767 601640
E-mail: eurospan@turpin-distribution.com

Cover design	Savitri Brant
Interior design	Donna Celso
Editing	Claudia Naber
Copyediting	Tiffany Wilkerson
Cartography	Riley Peake
Print production	Cliff Crabbe and Lilia Arias

On the cover: Courtesy of Suzano Papel e Celulose
Alex L. Fradkin/Stockbyte/Getty Images

Copyright retained by individual authors for all chapters except for Chapter 2: Imports and Exports of Roundwood in the Upper Midwestern United States, Chapter 6: Public and Private Forest Ownership in the Conterminous United States, and Chapter 18: Protecting the Drinking Water Supply in the Northeastern United States, which are provided courtesy of USDA Forest Service. No copyright is claimed in these chapters in the United States.

Contents

Introduction v

Mapping for business 1

Chapter 1	The feasibility of logging in the Pará Calha Norte region of the Brazilian Amazon	1
Chapter 2	Imports and exports of roundwood in the upper midwestern United States	5
Chapter 3	Modeling biomass transportation costs in North Karelia, Finland	9
Chapter 4	Competition for sawlogs in the Northern Forest	13
Chapter 5	Determining the stewardship potential of Indiana's nonindustrial private forests	17
Chapter 6	Public and private forest ownership in the conterminous United States	21

Mapping for inventory 25

Chapter 7	Analyzing urban forest characteristics in Florence, Alabama	25
Chapter 8	GIS for Romanian forest management planning	29
Chapter 9	Using a topographic index to define terrain types	33
Chapter 10	Analyzing the forest structure in northern Cambodia	37
Chapter 11	Using an integrated moisture index to assess forest composition and productivity	41

Mapping for operations 45

Chapter 12	Mapping land use at Swanton Pacific Ranch	45
Chapter 13	Improving watershed health and air quality in Washington, D.C.	49
Chapter 14	A wildfire risk management system for decision support	53

Mapping for sustainability 57

Chapter 15	Mapping southern pine beetle hazard in the Pisgah National Forest, North Carolina	57
Chapter 16	Predicting land-cover change and forest risk in the Bolivian lowlands	61
Chapter 17	Prioritizing restoration in fire-adapted forest systems	65
Chapter 18	Protecting the drinking water supply in the northeastern United States	69
Chapter 19	Improving sustainability planning in Brazilian eucalyptus forests	73

Author Information 77

Introduction

For more than one hundred years, professional foresters have been required to balance multiple resource objectives and constraints within the context of economic efficiency, productivity, and ecological sustainability. Contemporary forest managers still face such issues; however, now they must deal with unprecedented challenges posed by increasingly complex regulatory requirements and evolving social expectations. As a result, the business of sustainable forestry is currently undergoing transformation in an era of dramatic economic change and intensifying global competition. Geographic information system (GIS) software is helping foresters manage, view, and model data so they can meet environmental and commercial demands.

In many countries, GIS technology has had a profound impact on the way foresters manage the timber resource. Initially embraced to maintain more accurate timber inventories, GIS is evolving to become the foundation for new decision-support tools used in all areas of integrated forest management. Geography plays a role in nearly every decision made by a forest manager. The tasks of choosing harvesting sites, designing road networks, planning silviculture treatments, selecting harvesting equipment, and tracking invasive species involve answering geography-related questions.

This book gives readers a rare glimpse into a wide variety of regions throughout the world where geographic information software systems are used to support better forestry and land-management decisions. From developing countries to the developed world, forestry professionals rely on GIS to understand the timber resource, make strategic plans, manage resources sustainably, reduce costs, create efficiencies, and increase profitability. With the majority of chapters representing the application of GIS in the United States, a number of chapters also feature unique applications in Bolivia, Brazil, Cambodia, Canada, Finland, and Romania.

For those who work in forestry, but who are not GIS experts, this book can help explain the many benefits that GIS can contribute to a forestry organization. Each chapter introduces a situation then follows with "A Visual Solution" section that describes how GIS maps and analysis helped create a business solution. For GIS experts, this book can help you understand the workflows and data required to re-create maps for your own organization. The "Data Dictionary," "Software Dictionary," and "Recipe for Map-Building Success" sections will help GIS analysts understand the data requirements and analysis needed to produce similar results.

When using GIS to develop computer-based maps, field foresters, urban foresters, and forestry technicians rarely set out to develop high-quality cartographic products. Instead, forestry maps tend to be operationally focused with more attention placed on analysis and accuracy than on creating a map suitable for framing. However, this book aspires to highlight beautiful representations of forestry maps that have been developed for a full range of research, planning, and operational forestry activities.

Each map in this book represents many years of data collection undertaken by field specialists and geospatial experts who transform aerial photographs, satellite images, and data obtained from Global Positioning Systems (GPS) in a digital GIS environment. The overwhelming majority of maps was created using a variety of map layers from different sources.

All organizations included here have made the commitment to use GIS as a critical part of their decision-support infrastructure, and the descriptions of each of these maps clearly demonstrates a measurable return on their respective GIS investment. In every situation, GIS makes it possible to quickly make decisions with fewer resources. It is also worth noting that the analysis demonstrated could simply not have been completed cost-effectively with the kind of traditional analog methods used prior to the development of modern geospatial technologies.

While this book focuses on maps, it also explores something even more impressive: the systematic examination of various forestry challenges. By analyzing known spatial relationships using a set of rules and procedures, results are displayed in the form of a map, often with tables or charts to help explain and interpret the information presented.

The chapters in this book can be grouped into four categories. The first six chapters deal with how GIS supports a wide variety of business issues in today's forestry industry. Maps related to logging feasibility, import and export analysis, supply-chain optimization, the pressures of competition, maximizing resource utilization, and more are presented. The following five chapters describe how GIS is used for inventory in a variety of forest settings, including the urban environment. Maps about maintaining a forest inventory for forest management planning, analyzing forest structures, and predicting forest structure are included. The next three chapters focus on how GIS is used in forestry operations, again including a chapter on urban forest planning. The last five chapters cover mapping for sustainability. Improving forest health; understanding changes related to harvesting, fire, and pests; and improving sustainability planning are discussed in these chapters.

Foresters and land managers use GIS to create information products that help commercial forestry organizations and government agencies run better. At a time when organizations are being pressured to do more with less, GIS has saved hundreds of millions of dollars through increased productivity and efficiencies. With GIS, forestry organizations can realize a significant competitive advantage by having tools to quickly make better-informed and more reliable decisions.

CHAPTER 1

The feasibility of logging in the Pará Calha Norte region of the Brazilian Amazon

Carlos Souza Jr., Amintas Brandão Jr., Marco Lentini

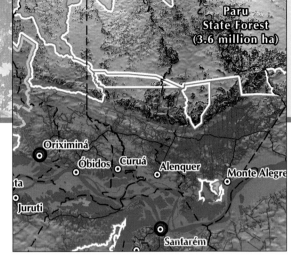

Throughout Brazil, forestry professionals are developing sustainable land management strategies to meet diverse stakeholder requirements. To manage and protect its forest resources, the state of Pará commissioned an analysis of the newly designated state forests and conservation units in Calha Norte. Rich in biological diversity, this region in the Brazilian Amazon encompasses an area of 12.8 million hectares, or 31.6 million acres, on the northern side of the Amazon River. The Calha Norte region borders with French Guiana, Suriname, Guyana, Venezuela, and Colombia.

Imazon, a Brazilian nongovernmental agency, has identified where logging is economically viable and where forests are at risk from illegal harvesting. It created a logging feasibility map that is used by regional stakeholders such as regulatory agencies, indigenous and local communities, logging companies, and conservation groups to negotiate and validate forest management decisions.

To create this map, Imazon first identified transportation networks, including roads and navigable rivers. Thematic layers such as topography, deforestation areas, conservation zones, and processing facilities were then combined to prepare the data for further analysis. Imazon GIS specialists stored the map data as rasters, which splits a map into a matrix of cells. Within a raster, each cell contains a value for such things as vegetation, slope, or soil type. The raster layer was then modeled using a least-cost-surface analysis technique to estimate the cost of transporting logs to the region's processing facilities. Understanding the least-cost surface helped determine where logging companies could harvest timber without incurring significant transportation costs and where illegal deforestation is most likely to occur.

A visual solution

In December 2006, the government of Brazil defined and designated State Forests and Conservation Units to begin the process of protecting national forest resources. By creating zones for tourism, community use, timber harvesting, and conservation, the region can be better managed by applying principles of sustainable forestry. The state government of Pará faces the challenge of both regulating the forest industry and protecting the region's rich biodiversity.

As the regional forest zoning process continues, the foresters need to understand the cost of transporting harvested timber. The logging feasibility map assists in valuating timber concessions and ensures a cost-effective sustainable timber supply for local manufacturing facilities. Additionally, the map helps enforcement officials identify areas at risk from illegal harvesting and other areas with high biodiversity levels that need protection.

By combining logging feasibility and protected areas in one map, forest managers can identify the extent of logging pressure within and around protected areas. With this knowledge, foresters can consider the entire landscape as an operating area as opposed to individual concession areas. This also provides the opportunity to develop reduced-impact logging allocations in some of the protected areas while still ensuring sustainable harvesting.

The feasibility of logging in the Pará Calha Norte region of the Brazilian Amazon

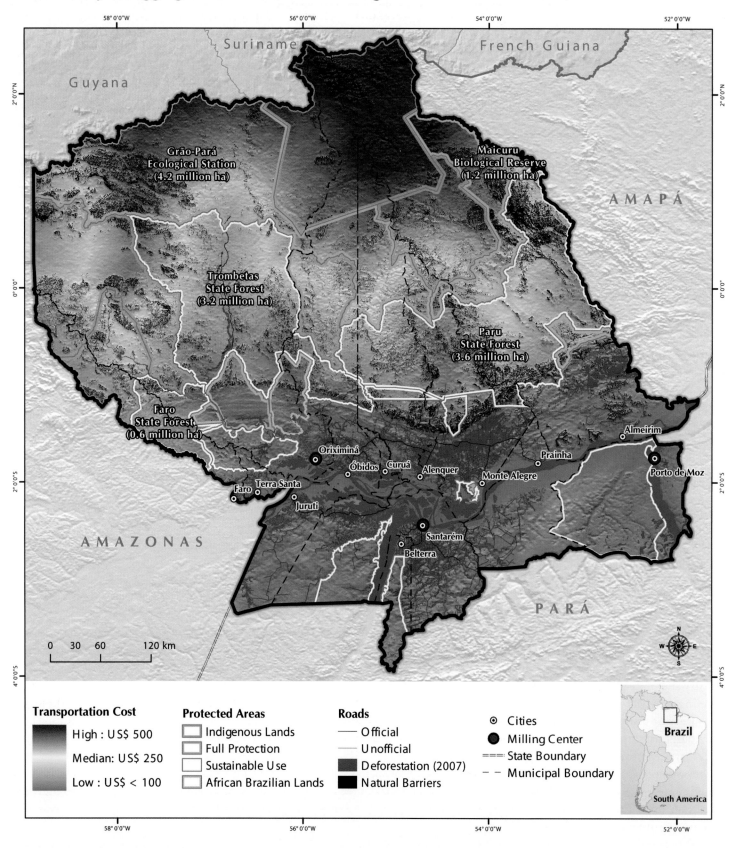

Data courtesy of Instituto Brasileiro do Meio Ambiente e dos Recursos Natruais Renovaveis (IBAMA); ISA; The National Aeronautics and Space Administration (NASA) and The National Imagery and Mapping Agency (NIMA); State Fundiary Institute of Pará (ITERPA); The Brazilian Institute of Geography and Statistics (IBGE); Amazon Institute of People and the Environment (Imazon); National Institute for Space Research (INPE).

Making the logging feasibility map requires technical experience with GIS. However, it is easy for nontechnical stakeholders to interpret the map, which clearly illustrates where logging activity is most profitable and where it will most likely occur. Imazon has been using these maps in public hearings to raise awareness and build consensus between enforcement officials, regulatory agencies, loggers, conservation groups, local and indigenous communities, and others.

Indigenous and African-Brazilian communities inhabit this region and, for the most part, depend on the forests for their livelihood. This map identifies local communities that are under pressure to cede forested land due to commercial harvesting activities. Therefore, this is also a powerful map to show where logging in and around these communities is prohibited by law. This information helps prevent social conflicts by communicating where deforestation is most likely to occur and allowing appropriate enforcement resources to be deployed.

Besides facilitating communication with stakeholders, this map is also used as input into a multicriteria decision-making tool developed to define the best areas for logging and community use and areas that should be kept as legal reserves due to rich biodiversity.

Resource tables

Data dictionary

General data description	Data sources
International, state, and municipal boundaries	Brazilian Institute of Geography and Statistics (IBGE).
Cities	Brazilian Institute of Geography and Statistics (IBGE).
Deforestation and rivers up to 2006	PRODES project.
Roads	Imazon.
Shuttle Radar Topography Mission (SRTM) data	Shuttle Radar Topography Mission.
Natural barriers	Spatial analyses based on SRTM data.
African-Brazilian land	Pará's Institute of Lands.
Protected areas: indigenous land	Instituto Sócio Ambiental.
Protected areas: full protection and sustainable use	Instituto Sócio Ambiental.
Study area delimited based on municipal boundaries	Brazilian Institute of Geography and Statistics (IBGE).
Milling centers	Imazon field survey.

Software dictionary

Software	Description
ESRI ArcGIS Desktop	Used to build the database, model the economic feasibility of logging, and generate maps and statistical reports.
ArcGIS Spatial Analyst extension	Used for raster-based spatial modeling and analysis.

Additional resources

Resource	Description and source
Logging socioeconomic database	Timber price and log transportation cost (source: Lentini, M., D. Pereira, D. Celentano, and R. Pereira. 2005. *Fatos Florestais da Amazônia 2005*. Belém: Imazon, 141).

Recipe for map-building success
The following steps outline how to create a similar logging feasibility map.

Step 1: Determine the scope of the project
The first step in any GIS project is to clearly understand the project objective. The goal of this project was to estimate the economic feasibility of logging in the twelve million hectares of the Calha Norte region of the state of Pará. Imazon used the least-cost-surface model, primarily because it is a scientifically validated process. To generate the least-cost-surface model, Imazon identified the required data layers, which included detailed information on roads, navigable rivers, natural barriers, vegetation types, deforestation, milling center locations, topography, transportation costs, and commercial timber species pricing.

Step 2: Procure and test data
After the data layers were identified, Imazon attempted to procure the data from a variety of sources. For data that wasn't readily available, such as for the roads, Imazon's senior remote sensing specialist and a team of GIS professionals generated the dataset by digitizing Landsat images, then visually interpreting those images.

Navigable rivers were defined by combining maps of rivers with information on waterfalls, rapids, river navigation reports, and Shuttle Radar Topography Mission (SRTM) topographic data. All nonnavigable rivers were classified as natural barriers that cannot be used for the purpose of transporting cut timber.

High-altitude areas were identified as natural barriers areas using a 3x3 moving window kernel with the block statistics function in ArcGIS.

The vegetation map was produced by classifying Landsat images and Japanese Earth Resources Satellite (JERS) images. Available coarse-resolution vegetation maps were used to define the vegetation types. Deforestation and other types of nonforest classes (i.e., savannas, natural fields, among others) were also used as possible types of surfaces to build roads to access forested regions.

Step 3: Build cost friction layer
Analysts generated a cost friction layer by combining maps of roads, navigable rivers, vegetation, and natural barriers into four independent layers using map algebra operations within the Spatial Analyst environment. The cell size of the cost friction layer was 100 meters.

From the cost friction layer, costs associated with transporting timber were then assigned to each surface type (US$/m^3 of raw timber/0.1 km) available in this map. These cost values were defined using empirical data collected during socio-economic field surveys (see table 1.1), including timber species prices, transportation costs, volume of timber harvested, harvesting locations, and more. To gather survey information, Imazon conducted interviews with loggers.

Step 4: Calculate path distance
Analysts used the path distance function with the cost friction map, the map of manufacturing facilities, and the SRTM topographic map to generate a least-cumulative-cost-surface map. This map was then reclassified based on the maximum cost of transportation for three groups of species values, ranging from high to low, into a logging economic extent map, which is illustrated in this chapter.

Step 5: Design cartographic presentation
The logging economic extent map was then combined with maps of protected areas, political boundaries, rivers, roads, deforestation, and milling centers to make an informative cartographic presentation.

Conclusion
This map accurately depicts areas subjected to deforestation, often because of illegal logging. By combining map layers, we can predict the level of threat that the region faces from overharvesting. Moreover, stakeholders consulted during public hearings have been using this map to discuss the regional zoning options that must accommodate demands from the timber sector, the local economy, and conservation interests. Finally, this map was adopted in a multicriteria decision algorithm that Imazon developed to produce a zone map of the entire Calha Norte region.

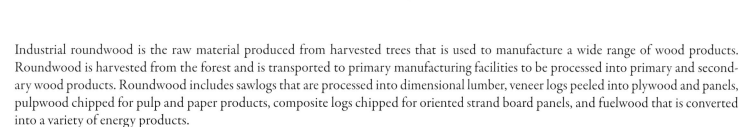

Imports and exports of roundwood in the upper midwestern United States

Charles H. (Hobie) Perry, Mark D. Nelson, Ronald J. Piva

Industrial roundwood is the raw material produced from harvested trees that is used to manufacture a wide range of wood products. Roundwood is harvested from the forest and is transported to primary manufacturing facilities to be processed into primary and secondary wood products. Roundwood includes sawlogs that are processed into dimensional lumber, veneer logs peeled into plywood and panels, pulpwood chipped for pulp and paper products, composite logs chipped for oriented strand board panels, and fuelwood that is converted into a variety of energy products.

This map of roundwood imports and exports in the upper midwestern United States illustrates the movement of roundwood out of state from harvest locations to primary wood proessing facilities. Roundwood processed in the same state where it is harvested is not depicted. The U.S. Department of Agriculture (USDA), Forest Service produced the primary map by drawing stylized vectors from harvesting operations to processing locations. We also produced two secondary maps (not shown here) that highlight those states with the largest volume of roundwood movement and identify net importers and exporters of raw logs.

We produced this map using mill survey data collected from primary wood processing facilities in the upper midwestern United States. We added each state's industrial roundwood receipts to a regional timber removal database and supplemented with data on out-of-state uses to provide a complete assessment.

A visual solution

The forest service's Forest Inventory and Analysis (FIA) program conducts forest inventories to enhance our understanding of the nation's forest resources. FIA data, information, and knowledge are collected, produced, and distributed to describe the biophysical, social, and economic benefits of forest resources for all types of timberland ownerships.

FIA's tabular data and summary reports are publicly available (see fiatools.fs.fed.us) and geospatial data and accompanying cartographic map products are becoming increasingly important to achieving FIA's mission. In this map we highlight FIA's economic product line by portraying the amount, distribution, and shipment of roundwood in the upper Midwest.

The general public often does not understand the complexity of timber flow. In particular, there is often confusion about the final destination of roundwood harvested in-state versus the overall source of wood processed at various mills throughout the region. This map helps the public understand the connectivity between forest resources, distant populations, and employment centers.

To help regulatory agencies and the forest products industry better understand the flow of roundwood, FIA designed a questionnaire to determine the source and destination of roundwood at the county level for all fifty states. FIA achieves close to a 100-percent response rate from primary mills through canvassing using mail questionnaires, phone surveys, and/or mill visits. As part of data editing and processing, all industrial roundwood volumes reported are converted to standard units of measure using regional conversion factors.

This map portrays interstate movement of roundwood based on survey data collected during the past decade from nearly 1,900 mills. The volume and direction of roundwood movements are portrayed using stylized vector features. Lines connect sources and destinations of roundwood, arrow symbols represent the direction of movement, line colors tie export routes to the state of origin, and line thickness represents the volume of roundwood.

Imports and exports of roundwood in the upper midwestern United States

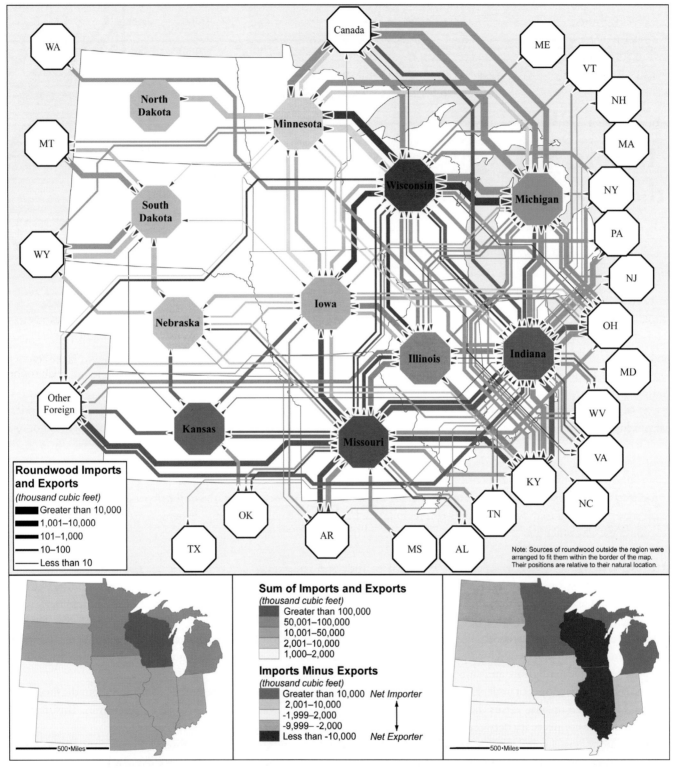

Data courtesy of ESRI Data & Maps, 2006, from ArcUSA, U.S. Census, ESRI (Pop2005 field); USDA Forest Service.

The number, volume, and complexity of import and export networks may be surprising. The map reveals that Indiana is by far the most diverse importer of roundwood, receiving roundwood from twenty-two different states and even some foreign countries. Wisconsin and Missouri are also diverse importers with fourteen different external sources of roundwood. The lake states of Michigan, Minnesota, and Wisconsin have the most timber flow, realizing more than 50 million cubic feet of annual imports and exports combined.

States can also be identified as net importers or net exporters of roundwood. Wisconsin and Illinois each export at least 10 million cubic feet of roundwood more than they import; Michigan and Minnesota, by contrast, are the largest net importers of roundwood.

Resource tables

Data dictionary

General data description	Data sources
Roundwood movements vector data layer	USDA Forest Service, Forest Inventory and Analysis Program.
Roundwood production point data layer	USDA Forest Service, Forest Inventory and Analysis Program.
Basemap layers of states	ESRI Data & Maps (See ESRI Web site).

Software dictionary

Software	Description
ESRI ArcGIS Desktop	Build vector data layer. Build point file for sources and destinations of roundwood. Layout and export map products.
Text editor	The original roundwood data files are available as simple text files. These files were reformatted to easily identify the origin and destination of each roundwood transfer. This data is joined to the vector layer created below.
	The text editor is also used to create the attribute files for the point data layers.

Additional resources

Resource	Description and source
FIA database documentation	Understand the format of FIA data and produce custom queries. fiatools.fs.fed.us
FIA's TPO analysts for consultation	North: Ron Piva, USDA Forest Service, rpiva@fs.fed.us
	South: Tony Johnson, USDA Forest Service, tjohnson09@fs.fed.us
	West: Todd Morgan, University of Montana, todd.morgan@business.umt.edu

Recipe for map-building success

Step 1: Develop and refine your question

An effective map addresses a specific question. Our original question was, what is the pattern of log movement from harvest sites to mills across the United States? A review of the available data made clear that this question is too complex to address in a small-format map. We refined our question to generalize our unit of interest to the state level, thereby simplifying the data by summarizing it at a coarser scale. We then restricted our field of interest to the Midwest, balancing a desire for detail with the constraints of the demonstrated map format.

Step 2: Identify the audience

Data collected on timber products output (TPO) commonly includes attributes that are economically sensitive, so published results must be presented in a way that protects confidentiality. Selecting states as the basic unit of interest achieves this end. Our intended audience includes timber buyers, mill owners, forest landowners, and the general public.

Step 3: Decide if a map is the best communication tool

Many different customers are interested in our field data collections and survey results. Our traditional users are accustomed to tabular summaries of TPO data. One of our goals for this map was to introduce a new, spatial perspective on timber products output.

Step 4: Acquire, understand, and prepare tabular data

The information collected by FIA is provided online at fiatools.fs.fed.us. Our databases are complex, so it is helpful to review the documentation and become familiar with available tables, attributes, states, and years of inventory. For this map, we acquired a text file that summarized the volume of harvested roundwood by state of origin and state of destination. The database was simplified to include only those transfers that included states in the upper Midwest. We added a unique field combining the states of origin and destination to join the table with the vectors created below.

Step 6: Create new spatial files

Three primary data layers were created for this map: (1) a series of points highlighting the "center" of each state, (2) a related series of points identifying necessary states outside the region, and (3) a series of vectors from roundwood sources to destinations. Each of these layers was joined with tabular attribute data such as state name and volume of roundwood in transit.

Step 7: Summarize the data

This data is extremely complex, and summaries improve the reader's understanding. We wanted to highlight two significant issues: (1) the total volume of roundwood in transit between states, and (2) the net balance of imports and exports.

Step 8: Prepare the map

An effective map includes several elements: scale bars, legends, neatlines, and consistent typography, to name just a few. Our program has defined cartographic standards to give each map a consistent look and feel.

Conclusion

Estimates of roundwood production and consumption are readily available at the county and state level, but few maps highlight the movement of roundwood between states. Our choice to represent roundwood movement as vector features offers a unique perspective on the movement of raw forest products. It also makes it possible to attribute and stylize the routes more efficiently than is possible with graphics software. ArcGIS facilitates the development of a spatial database of roundwood movement that can be adapted to the needs of individual states.

The style of cartographic illustration used in this map is unique in the industry. Traditional maps of timber procurement display actual routes from harvest sites to mill sites, but complete data for such maps is not available over larger geographic extents or for multiple owners of timberland resources. Readers of this set of maps readily observe the substantial and complex pattern of interstate roundwood movement. Movements of other resources and populations and have been portrayed in map formats that provide partial solutions for our needs; maps of waterfowl migrations show generalized flyways and relative population sizes.

We created the main map in a style that purposefully mimics public transportation system maps from major metropolitan areas. Transportation maps, like subway maps, show stylized routes, but don't capture directional dependencies or volumes of passengers. Used in a forestry context, this design is unique, yet familiar because of its cartographic lineage. The inset maps serve as a complement to the main map by simplifying and summarizing the seemingly chaotic pattern of movement. By integrating components from these and other maps, we were able to accomplish our objective of cartographically portraying several characteristics simultaneously, including sources, routes, directions, and volumes of roundwood movements between states and other countries.

Modeling biomass transportation costs in North Karelia, Finland

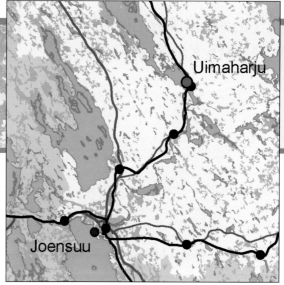

Perttu Anttila, Lauri Sikanen, Dominik Röser,
Juha Laitila, Timo Tahvanainen, Heikki Parikka

The desire for energy self-sufficiency and the European Union's binding targets for the use of renewable energy by 2020 to mitigate climate change motivates countries in Europe to increase their share of renewable energy alternatives. Biomass is one of the most important sources of renewable energy in Europe. In countries like Finland the use of wood-based energy production has a long tradition because of abundant forest resources and large-scale forest products manufacturing.

While political pressure encourages the increased use of wood as an energy source, biomass resources in Finland are not evenly distributed. In some areas the demand is exceeding supply, while in many others, opportunities exist to increase the volume of wood chips to produce even more energy.

Forest biomass is remarkably different than traditional energy sources like oil and coal since the energy density of forest biomass is much lower. Establishing supply chains and transporting wood fuel comprise the most significant cost factors. The location of energy conversion facilities, the preferred supply chain, and long-distance transportation options need careful consideration to ensure a sustainable and cost-effective supply of wood chips to each combined heat and power plant.

This map was developed by the Finnish Forest Research Institute (Metla) to illustrate the potential supply chains to transport forest biomass for energy from the province of North Karelia to the Uimaharju Combined Heat and Power (CHP) plant. Fifteen supply chains were initially analyzed to determine the lowest cost of delivered biomass to the CHP plant. Only the three lowest-cost supply chains are identified on this map for management consideration.

A visual solution

Abundant forests exist throughout Finland, but the spatial distribution of important factors such as tree species, age, technical quality, or growing sites vary considerably. This variation influences where and how to best use forest resources for both traditional lumber manufacturing and biomass extraction for energy production.

To generate heat and power from biomass, logging residues must first be chipped into small flakes then transported to a mill that converts the wood chips into energy. In North Karelia, spatial analysts discovered that there are three cost-effective methods of chipping and transporting biomass, which have been identified on this map. In the first chain, "Chip Truck" logging residue is collected, chipped at the roadside, and then transported to the CHP plant by truck. In the second chain, "Truck and Train" residues are also chipped at the roadside, but in this method chips are transported by truck to a nearby railway terminal and then by train to the plant. In the third chain, "Loose Residue Truck" logging residues are transported by truck to the plant, where they are then chipped.

When interpreting this map, it becomes evident that the cost of transporting wood chips varies spatially. This map illustrates the cheapest supply chains to the Uimaharju CHP plant. This information can then be used to manage and allocate appropriate corporate resources to ensure that the CHP plant has a continuous supply of biomass.

This analytical process can be applied within the context of a feasibility study to decide whether, and where, to construct a biomass power plant. With detailed data on biomass fuel resources, growing stock, future harvesting activities, and transportation options, the plant can be located and scaled correctly.

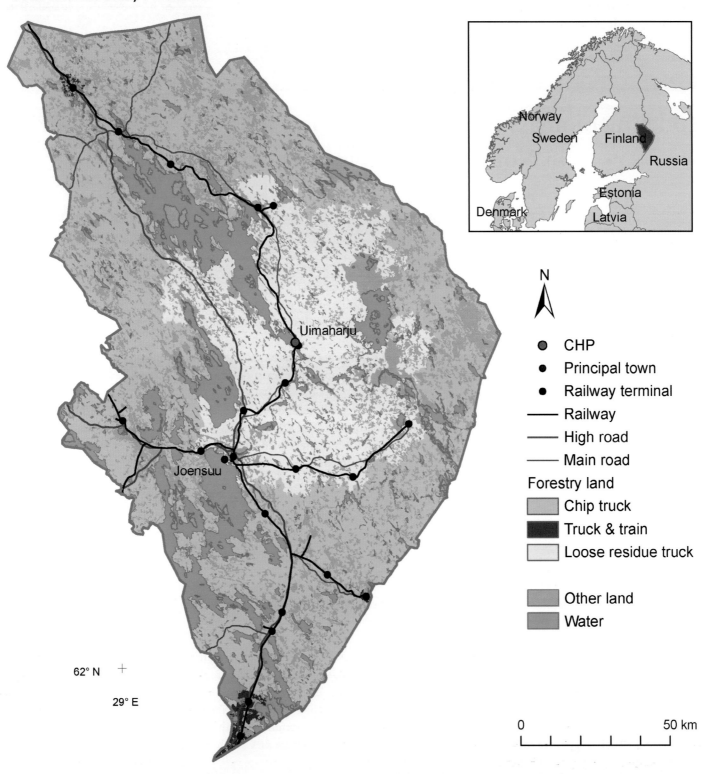

The map's yellow areas describe the regions where loose-residue trucking is the most effective transportation option. One can see that the area is not defined by a certain Euclidean distance around the mill, but the restrictive waterways and the limitations of the transportation network strongly affect the real supply cost. The analysis and map also help in strategic decision making. The map shows that train transportation is the best option only on relatively small areas in the north and south. Therefore, establishing a railway-based transportation system should probably be avoided unless procurement is extended to other provinces.

Resource tables

Data dictionary

General data description	Data sources
Province boundaries	National Land Survey of Finland. Dataset: Hallintorajat
Railways	National Land Survey of Finland. Dataset: Maastotietokanta
Roads	The Finnish Road Administration. Dataset: Digiroad
Railway terminals	The Finnish Rail Administration. Dataset: Rautatieliikennepaikat
Forest land	Finnish Environmental Institute. Dataset: CLC2000
Conservation areas	Finnish Environment Institute. Datasets: Luonnonsuojelualueet and Natura2000
Country boundaries	ESRI Data & Maps. Feature dataset: cntry06.sdc

Software dictionary

Software	Description
ESRI ArcGIS Desktop	Tools used: clip, erase, extract by mask, lowest position.
ArcGIS 3D Analyst extension	Interpolate to raster using the Natural Neighbors tool.
ArcGIS Network Analyst extension	Origin-destination (OD) cost matrix.

Additional resources

Resource	Description and source
Cost models for the supply chains	Used in calculating total costs of chips. Sources:
	Laitila, J. Cost and sensitive analysis tools for forest energy procurement chains. *Forestry Studies* 2006 (45):5-10.
	Ranta, T. 2002. Logging residues from regeneration fellings for biofuel production: A GIS-based availability and supply cost analysis. Doctoral thesis, Lappeenranta University of Technology, 180.
	Ranta, T., and S. Rinne. The profitability of transporting uncomminuted raw material in Finland. *Biomass and Bioenergy* 30 (2006):231-7.
	Korpilahti, A. 2004. Oksapaalien autokuljetus. Metsätehon raportti 169, p. 25. In Finnish.
	Anttila, P., T. Tahvanainen, H. Parikka, J. Laitila, and A. Ala-Fossi. 2007. Energy wood transportation by rail.
	Joensuu. Metla. 5 EURES (EIE/04/ 086/S07.38582). Project report 23 — Case study of North Karelia, 18.

Recipe for map-building success
Metla used the following steps to model biomass transportation costs:

Step 1: Determine objective

We determined the objective of the map based on the type of strategic and tactical decisions needed, then identified the geographic area of interest.

Step 2: Identify supply chain options

We analyzed chains that are currently in use and other possible chains. However, we selected no more than five chains at a time to clearly understand the results.

Step 3: Collect and prepare data

We obtained the following data for the analysis: plant location, forest data, roads, railroads, terminals, and costs. **Tip:** Sources of the data can be land surveys, commercial map and remote sensing data providers, national forest or natural resource inventories, universities or research and development organizations, local authorities, forest companies, and organization networks. Since investment decisions will be made based on the analysis, ensure that the data is of high quality.

We then prepared the data for the analysis. **Tip:** Often, procuring and preparing data takes more time than the actual analysis. Clip the datasets with the outer boundaries of the region of interest. Verify that a coordinate system is defined for all layers. Protected areas such as riparian buffers, parks, and ecological reserves are removed from the forest land layer. Create a network dataset of the roads and railways.

Step 4: Calculate origin-destination cost matrix

a. First, we modeled supply chain costs to create optimal supply regions for each of the supply chains. We based costs on our research on Finnish logging activities, including forwarding, chipping, and transporting logging residues. In addition, we used hourly machine cost data and yields of logging residues from historical harvesting data in the calculations.

b. We then calculated origin-destination cost matrix for each forest site and plant. The cost matrix included the distance, driving time, and speed for each road section. **Tip:** If specific costs are not available, formulate cost functions by estimating the total cost of transporting the chips to the plant by summing the costs associated with chipping, hauling, transportation, and so on. These, in turn, can be broken down by working time and hourly costs.

c. We then calculated total costs for chips delivered from each forest site with each supply chain.

Step 5: Interpolate cost surfaces

We interpolated a cost surface for each chain with initial point-level data using ArcGIS Spatial Analyst's Natural Neighbor tool. Subsequently, we used the Extract by Mask function to exclude nonforestry land. The forest cover layer was generated by combining CORINE (Coordination of Information on the Environment) land-cover data (see reports.eea.europa.eu/COR0-landcover/en) and data on conservation areas that are restricted from harvesting. Finally, we determined the cheapest chain in each pixel by using the Lowest Position command on all cost surfaces. Then we were able to create a layer of supply regions.

Step 6: Create map

We added cartographic elements to our map such as a title, index map, legend, scale bar, north arrow, coordinate grid, explanatory texts, and logos.

Conclusion

The objective of this map was to visualize the extent and costs of supply areas using a variety of supply chain options. While the objective was met, as demonstrated by the areas clearly identified on the map, the validity of the analysis was completely dependent on the quality of initial data: land cover classification, road and rail networks, and cost models. We believe that the map reflects the real world successfully because the models are based on significant research. Additionally, we were able to use detailed and accurate Finnish road network and CORINE land cover classification data.

We attempted to minimize the number of map layers to ensure that the regions were clear and obvious to the reader. We feel that the quality of the data and the knowledge of supply technology behind the map make it especially successful.

Metla wanted to show and share our expertise related to the technology of harvesting energy wood. The map illustrates only one option to support biomass recovery. Metla has also created detailed supply analyses at the stand level. Every forest stand has unique values for biomass potential per hectare based on various harvesting conditions. These maps can be combined with the transportation option map, but then the map itself becomes increasingly difficult for nonforestry professionals to interpret.

Competition for sawlogs in the Northern Forest

Nate Anderson, Eddie Bevilacqua, René Germain

The Northern Forest includes 10.5 million hectares (26 million acres) of contiguous mixed hardwood and coniferous forest that stretches across New York, Vermont, New Hampshire, and Maine. This mosaic of state, federal, and private land provides a wealth of resources and ecosystems, including critical habitat and water resources.

Since the early 1800s, timber from the Northern Forest has been the foundation of a vibrant lumber industry, which remains a source of economic stability for many of the region's communities. The forest industry in the four Northern Forest states provides $14.4 billion in manufacturing shipments that account for 7.3 percent of all manufacturing sales. In rural communities, this industry is responsible for over 90,000 high-paying manufacturing jobs.

In 2004, researchers at the State University of New York (SUNY) College of Environmental Science and Forestry began a study focused on quantifying and predicting the effects of land-use change on the Northeast sawmill industry. Because most mills purchase roundwood on the open market as stumpage or logs, it is expected that land-use changes negatively affecting the sawlog supply will be felt most quickly and discernibly where competition for the resource is most intense.

A visual solution

Sawmills in the northeastern United States tend to compete with other mills for their wood supply rather than owning or leasing their own timberlands. Since log exports are a major component of log shipments from the Northern Forest, U.S. mills in this region must also compete with companies located in Canada and abroad. Some large Canadian companies procure more than 90 percent of their total sawlog supply from the United States; yet in contrast, few American companies in the Northern Forest use imported logs.

A mail survey sent to more than 700 sawmills in the United States and Canada provided the data for this analysis. Given that log procurement is typically the dominant cost of production for sawmills, understanding the spatial dynamics of competition for this resource can have tangible economic benefits.

In addition to relevant procurement information, representatives from each mill identified the geographic extent of their wood procurement region, also known as a woodshed. This project represents the first research activity in the region to provide a landscape-level visualization of competitive pressure based on a synthesis of woodshed maps provided by individual mills.

The map creates an empirical framework for testing hypotheses and predicting changes related to the many variables that affect wood supply, including industry concentration, lumber market conditions, currency fluctuations, and land-use and ownership changes. It identifies hot spots of competitive pressure and serves as the foundation of geospatial models linking urbanization to the future timber supply in this area.

To assess these effects geographically, specific areas with competitive sawlog markets were mapped using data provided by sawmills. The map visualizes the level of competition for sawlogs in 2005 and shows all 291 sawmills in the study area, the intensity of woodshed overlap, and the baseline competition index in thousand board feet (mbf) per square mile. The baseline competition index represents the weighted sum of the overlapping woodsheds.

Competition for sawlogs in the Northern Forest

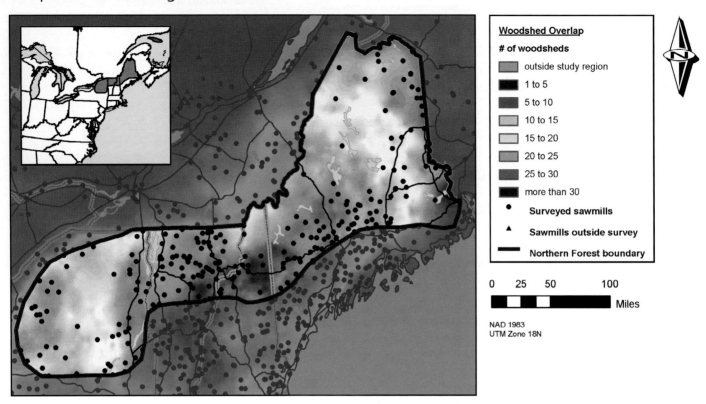

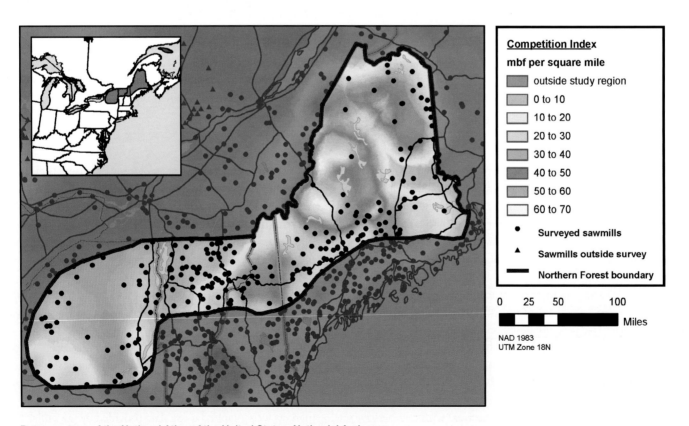

Data courtesy of the National Atlas of the United States; Nathaniel Anderson.

This project also has important applications for predicting changes in the composition and structure of the Northeast's forests. In general, timber harvesting is a dominant disturbance in these forest ecosystems. Areas with high competition for sawlogs are likely to experience higher harvesting pressure and greater disturbance. We can test such hypotheses by examining the spatial relationships between competition and other variables, such as forest stocking and removals. For example, across the Northeast, the urbanization of forestland has the potential to increase the costs associated with log procurement by both decreasing the amount of forest that is available for timber harvesting and by reducing the quality and productivity of remaining timberland.

Resource tables

Data dictionary

General data description	Data sources
Basemap data (including boundaries, roads, and other features)	National Atlas
Sawmill locations	USDA Forest Service, Southern Research Station. File: Northeastern U.S., mills_ne.zip
	Global Forest Watch Canada. File: Canada's Forest Product Mills, gfwc_mills_2004
	Sawmill industry directories published by state agencies.
	The Atlas of Canada, Natural Resources Canada.
	Spelter, H. and M. Alderman. 2005. Profile 2005: Softwood sawmills in the United States and Canada. Research Paper FPL-RP-630. Madison, Wisc.: USDA Forest Service, Forest Products Laboratory.
Woodshed overlap and competition index rasters	2006 SUNY-ESF sawmill survey.

Software dictionary

Software	Description
ESRI ArcGIS Desktop	Used to build the database, model the economic feasibility of logging, and generate maps and reports.
ArcGIS Spatial Analyst extension	Used for raster-based spatial modeling and analysis.
ArcGIS Geostatistical Analyst extension	Used for statistical analysis.

Additional resources

Resource	Description and source
ArcScripts	Split layer by attributes.
	Created in Python by Dan Patterson.
	SplitLayerByAttributes.zip
	arcscripts.esri.com/details.asp?dbid=14127
Personnel	Personnel should be familiar using both vector and raster data techniques, including heads-up digitizing, model building, and spatial analyst tools.

Recipe for map-building success

To make this map, graduate students received introductory and advanced ArcGIS software training and courses in applied geostatistics.

Step 1: Collect data

Through a mail-in survey, mill managers delineated geographic areas that identified woodsheds for their respective sawmills. The survey had a response rate more than 50 percent, representing 1.2 billion board feet of procured sawlogs or about 75 percent of the total sawmill wood procured in the study region. This is a strong response rate for a survey of this kind, but collecting data in this way is costly and time-consuming.

Step 2: Digitize woodsheds and join procurement data

Woodsheds sketched on analog maps were converted into a digital format using a digitizing tablet and on the computer monitor in the heads-up editing environment. The newly created polygons were then joined to mill data collected during the survey.

Step 3: Calculate weighting factors

Weighting factors were determined based on combinations of variables; these factors would provide the foundation for landscape-level evaluations. All woodsheds were equally weighted to provide a response surface that reflects the intensity of woodshed overlap, as demonstrated on the top map. Annual procurement volume per square mile of woodshed area (mbf/mi^2) is the weighting factor used in the extent pictured on the bottom map. This approach is used to examine such characteristics as competition for hardwood stumpage or the overall intensity of timber leasing.

Step 4: Rasterize and sum

Individual woodshed polygons were converted to rasters in order to take full advantage of the tools available in the ArcGIS Spatial Analyst extension and a wide range of geostatiscal functions, including mathematical operations such as weighted sum. We used the freely available Split Layer by Attributes script to isolate individual polygons or groups of polygons for conversion to raster format. This script ensures that the weight of interest becomes the cell value for each woodshed, and the sum of all woodshed rasters can be masked to the study region.

Step 6: Perform raster smoothing

Overlapping the sampled woodsheds can produce noisy edge effects near political and landscape features that are natural woodshed boundaries for many mills in the same area. We used a moving average filter to smooth the response surface where such effects occurred.

Step 7: Symbolize and map

We visualized the resulting rasters in various ways to demonstrate the effectiveness of the analysis and distribute the results to other researchers and industry representatives. We also used the data as an input to further geostatistical analysis and scientific research activities.

Conclusion

Though forestry and economics research scientists have been studying the spatial dynamics of markets for some time, GIS provides a powerful tool for identifying and quantifying spatial patterns in resource procurement, including procurement of pulp, biomass, sawlogs, veneer, and other products used by the primary forest products industry. Unlike previous research that relies on theoretical woodshed boundaries, this map is based on maps of woodsheds provided directly by individual sawmills. In addition, based on the data and maps provided by these sawmills, researchers modeled the woodsheds of nonrespondent mills to provide an analysis that includes all mills in the study region. To the degree that it captures the true geographic extent of procurement operations for the sampled mills, this approach represents an improvement over previous methods and opens the door to more accurate inference.

The map featured here also provides a clear example of the applied nature of combined geospatial and market research activities. For industry professionals, it is important to understand the hot spots of competition since these can translate into higher prices for roundwood, which is the most costly component of lumber production. In addition to industry applications, maps like this one can be used to clearly and quickly demonstrate spatial patterns to policy makers who make important management decisions that impact the log supply and the forest products industry.

While this map was designed to be accessible to forest industry stakeholders, it forms the foundation for increasingly complex yet meaningful geospatial analyses that focus on evaluating the effects of land-use change in relation to the future of log supply in this region. The ability to visualize such analyses on a map, rather than through the interpretation of statistical tables makes this research both more accessible and increasingly influential.

CHAPTER 5

Determining the stewardship potential of Indiana's nonindustrial private forests

Andriy V. Zhalnin, Richard L. Farnsworth,
Shorna Broussard Allred, Brett A. Martin

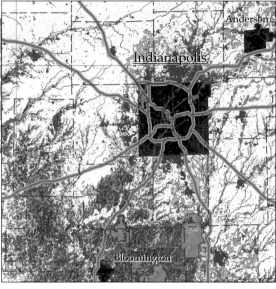

In the United States, private citizens own approximately 260 million acres of forests, which equals nearly 42 percent of the nation's forests. To improve management of these nonindustrial private forests (NIPF), the U.S. Department of Agriculture (USDA), Forest Service created the Forest Stewardship Program. This program helps NIPF landowners develop comprehensive, multiresource management plans needed for long-term forest management. The Forest Stewardship Program is administered by the forest service and implemented by state forest agencies. The program has grown steadily since 1990 and now helps with more than 29 million acres of land under professional forestry management.

Forest Stewardship Potential is the title of a three-color map that depicts the suitability of Indiana's privately owned forests for enrollment in the Forest Stewardship Program. Analysts used a combination of twelve GIS maps representing the major resources and threats to those resources to estimate stewardship potential. The nine resource-rich maps—riparian areas, priority watersheds, forest patch size, natural heritage data, public drinking water supply sources, private forest lands, proximity to public lands, wetlands, and topographic slope—capture landscape characteristics that contribute to the flow of forest products and related ecosystem services. The three resource-threat themes—forest health, housing density, and fire potential—are stressors, contributing to the reduction of forest products and services. Analysts weigh themes equally or unequally according to how forestry officials deem their importance. Forests in the high category are rich in resources and vulnerable to natural or human-induced threats, making them prime candidates for enrollment into the program.

A visual solution

A team of forestry and GIS professionals from the forest service, Maryland, Connecticut, Massachusetts, and Missouri created the Spatial Analysis Project (SAP) to provide a consistent approach for understanding and then improving implementation of the Forest Stewardship Program. At the core of SAP is a GIS-based methodology that allows every participating state to estimate and construct a three-colored map of stewardship potential of private land within its borders.

Once completed, a state's forest stewardship potential map can serve three purposes. First, states can use the map to identify hotspots—high stewardship potential areas—within their borders. Second, using the map and its underlying data, states can overlay enrolled stewardship acres and determine if the existing Forest Stewardship Program is effectively enrolling "high" stewardship acres. Third, states may find their current allocation of funds and staff at odds with the existing distribution of stewardship potential lands, thus creating the stimulus to reorganize.

Indiana's Division of Forestry joined SAP in 2004. It contracted with Purdue University researchers to collect the required spatial data, apply the SAP methodology, create the map of stewardship potential according to SAP guidelines, and digitize the state's forest stewardship plans. Thus far, Indiana's state forester has used this information to identify highly valued private forestland within the state, redraw district boundaries, reassign staff to try to increase enrollment of high stewardship potential lands in the Forest Stewardship Program, and assess the stewardship potential of forestland already enrolled in the program.

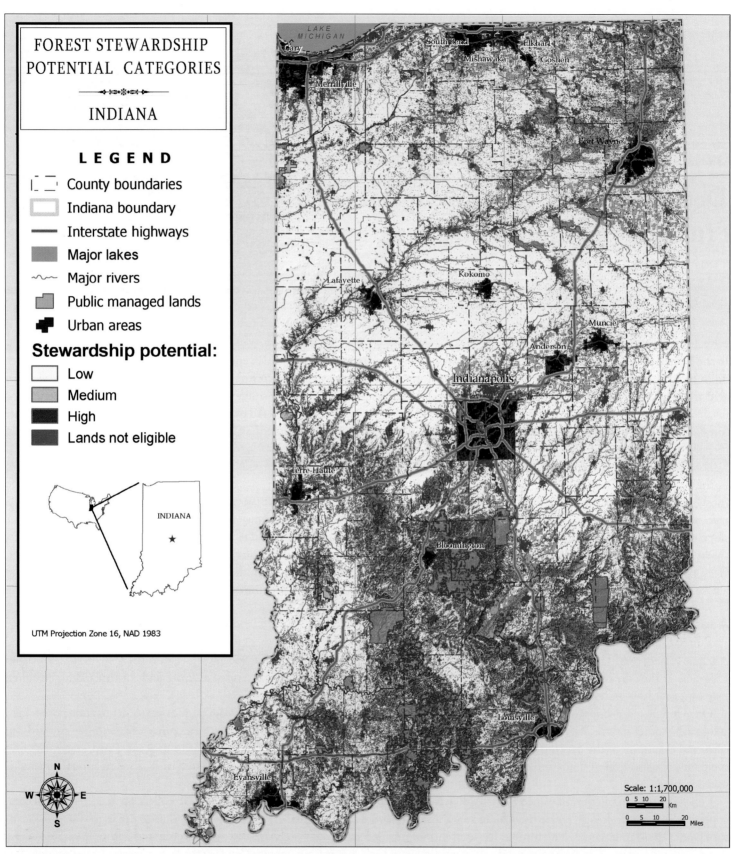

Data courtesy of USDA Forest Service; the National Atlas of the United States, U.S. Geological Survey; Indiana Natural Heritage Data Center, Indiana Department of Natural Resources; the U.S. Department of Commerce, Census Bureau; U.S. Geological Survey, EROS Data Center; U.S. Fish and Wildlife Service.

Resource tables

Data dictionary

General data description	Data sources
Map of forest stewardship potential categories	USDA Forest Service; Indiana Department of Natural Resources—Division of Forestry, Indianapolis, Indiana. See the USDA Forest Service Forest Stewardship Spatial Analysis Project Web site.
Map of forested land in Indiana	U.S. Geological Survey National Land Cover Database (NLCD 2001). See the Multi-Resolution Land Characteristics Consortium Web site.
Map of forest pest infestation potential, Map of wildfire risk	Indiana Department of Natural Resources, Indianapolis, Indiana.
Map of public water supply	U.S. Geological Survey and Indiana Department of Environmental Management.
Map of risk of development	USDA Forest Service, North Central Research Station, Forest Inventory and Analysis Program.
Map of impaired watersheds	U.S. Geological Survey and Indiana Department of Environmental Management.
Map of wetlands	U.S. Fish and Wildlife Service, National Wetland Inventory.
Map of public managed lands	Indiana Department of Natural Resources, Division of Forestry, Indianapolis, Indiana.
Map of urban areas, Map of roads	U.S. Census Bureau.
Map of streams and lakes	U.S. Geological Survey National Hydrography Dataset, available on the USGS Web site.
Map of shaded relief	U.S. Geological Survey National Elevation Dataset, available on the USGS Web site.
Map of Indiana state, Map of the United States	U.S. Department of the Interior. See the National Atlas Web site.

Software dictionary

Software	Description
ESRI ArcGIS Desktop	ArcToolbox, ArcEditor, ModelBuilder.
ArcGIS Spatial Analyst extension	Used for raster-based modeling and analysis.
ArcGIS ArcCatalog	Manage layers and develop metadata.
ArcGIS Maplex extension	Extend labeling tools.

Additional resources

Resource	Description and source
A. V. Zhalnin, S. Broussard, and R. L. Farnsworth.	Zhalnin, A. V., S. Broussard, and R. L. Farnsworth. 2008. Effectiveness of forest stewardship program in conserving natural resources on private lands in Indiana, USA. In: D. F. Jacobs and C. H. Michler, eds. Proceedings, 16th Central Hardwood Forest Conference. April 8-9, 2008, West Lafayette, Indiana. General Technical Report NRS-P-24. Newton Square, Penn.: USDA Forest Service, Northern Research Station, 314-22.

Recipe for map-building success

Step 1: Data collection and preparation

Because we were using SAP methodology, first we developed twelve maps differentiated into two groups:

a. Resource potential factors: riparian zones, priority watersheds, forest patch size, natural heritage data, public drinking water supply sources, private forest lands, proximity to public lands, wetlands, and topographic slope.
b. Resource threat factors: forest health, land development level, and wildfire assessment.

All maps consisted of polygons that display the specific factor. The riparian areas map, for example, included only 300-foot buffered areas adjacent to streams.

Step 2: Weighting scheme development

The relative importance of each of the twelve map themes must now be determined. Members of Indiana's Forest Stewardship Coordinating Committee and Indiana Department of Natural Resources district foresters were asked to weight the maps according to their relative importance. Each person was given twenty dots and instructed to distribute these dots among the twelve maps hung around the conference room. No person could put more than five dots on a map. Dots per map were divided by total dots to weight the individual maps. Participants quickly grasped this approach, agreed with it, and supported the final weighting scheme.

Step 3: SAP analysis

The geospatial analysis was done with the ArcGIS Spatial Analyst extension, which requires that all vector data be converted to a raster grid data format. Masks were created for areas that needed to be excluded from the analysis, and then ArcGIS ModelBuilder was used to do the following:

a. Reclassify each pixel in the individual resource maps into binary-type code. Where resources are absent, reclassify to "0" and where resources are present reclassify to "1".
b. Apply the weighting scheme previously developed to each resource layer.
c. Add all layers and convert pixel values from "0-12" scale (for twelve layers) to "0-1" scale that represents stewardship potential with a value of "1" representing the highest level of suitability.

We determined thresholds for three categories (high-medium-low) using the natural break classification scheme and created a final aggregated and categorized map. The algorithm classifies the data on natural groupings inherent in the data and picks class breaks that best group similar values and maximize the differences between classes.

Step 4: Production of the final map

We used Spatial Analyst's hillshade function to generate a shaded relief background map from the digital elevation model to enhance the visual appeal of the map. We added other natural and man-made features from the landscape (e.g., cities, roads, streams, lakes, etc.) to make the map easier to navigate and included additional information about stewardship potential categories and how they were created to help people better understand the map. We used the ArcGIS Maplex extension to label features on the map.

Conclusion

Millions of enrolled acres and thousands of applied forestry practices clearly indicate that the Forest Stewardship Program is popular among private forest owners. Program success, however, needs to be measured against the program's objective of enrolling and managing economically valued forests. The Spatial Analysis Project and its primary product, a map of forest stewardship potential, represent a step in the right direction for several reasons. First, location plays a critical role in the amount and quality of forest products and related ecosystem services that flow from a forest parcel. The GIS-based methodology used to estimate stewardship potential directly accounts for location effects.

Second, the use of map themes to characterize resource richness and resource threats significantly reduces data requirements, incorporates location into the analysis, and makes it easy for users to view and understand the individual factors that contribute to scoring and grouping stewardship potential. The ease in which users grasp the linkages between map themes and resource richness and threats makes later tasks such as deciding whether to add more themes and weighting the relative importance of the themes much easier.

Third, and most important, the three-color Forest Stewardship Potential map is not only easy to view and understand, but it also sets the stage for assessing program effectiveness and future program changes. In a few minutes foresters and the public can view the mix of low-, medium-, and high-stewardship-potential lands in their state and identify hot spots. The next activity would be adding the digitized Forest Stewardship Program contracts, district forest boundaries, and staffing plan map layers to the stewardship potential map to assess existing program effectiveness and possibly realign staff and funding to hot spots.

Public and private forest ownership in the conterminous United States

Greg C. Liknes, Mark D. Nelson, Brett J. Butler

Forests and the goods and services they provide are influenced by both the biophysical and human environments. To fully understand forest ecosystems, we need to understand the social context in which forests exist because landowners determine land use and management practice. To influence decisions related to the forests, we need to understand the spatial distribution of forest ownership.

The U.S. Department of Agriculture (USDA), Forest Service's Forest Inventory and Analysis (FIA) program collects annual information on the status, health, and trends of forests across all land ownerships. As the nation's forest census, the FIA program develops map products on the social, biophysical, and economic state of our forests. Further, FIA's National Woodland Owner Survey gathers additional information on forest owners' demographics, values, concerns, intentions, and uses of their land.

The maps in this chapter depict public and private forest landownership in the conterminous United States. They were produced by combining forest type maps from the forest service with a protected areas database from the Conservation Biology Institute. In the upper map, private forests are further stratified by corporate ownership, using data from the Resources Planning Act (RPA) forest resource assessment. In this context, forest land is defined as land which is at least 10 percent stocked by forest trees of any size, including land that had such tree cover that will be naturally or artificially regenerated. Forests owned by an incorporated business are referred to as corporate forest land. Examples of incorporated businesses include integrated forest products companies, timber investment management organizations, and real estate investment trusts.

A visual solution

Misperceptions about the ownership of forests in the United States are widespread. A recent poll indicates that 60 percent of voters think forests are predominantly government-owned, and an additional 20 percent think forests are predominantly owned by forest industry companies. In reality, of the 751 million acres[1] (304 million hectares) of forest land in the United States, 56 percent (421 million acres or 170 million hectares) is privately owned. Only one-third of this privately owned land is held by corporations. A better understanding of the characteristics and distribution of forest ownership enables researchers, politicians, landowners, and interested citizens to make more informed decisions about the nation's natural resources. This map helps correct the misperceptions about forest ownership in the United States and highlights important regional differences. The maps and similar products are used in national, regional, and state-level forest assessments aimed at understanding current patterns and future trends in forest resources.

In the upper map, the percentage of private forest land in corporate ownership is summarized across a hexagon sampling array, which provides spatial information while maintaining landowner privacy. The lower map shows the spatial distribution of hardwoods and softwoods, with corresponding bar charts comparing the amount of forest land area (millions of acres) in eastern and western United States by ownership and forest type categories.

[1] Smith, B., et al. 2009. Forest Resources of the United States, 2007. WO-78. Washington, D.C.: U. S. Department of Agriculture, Forest Service, Washington Office.

Public and private forest ownership in the conterminous United States

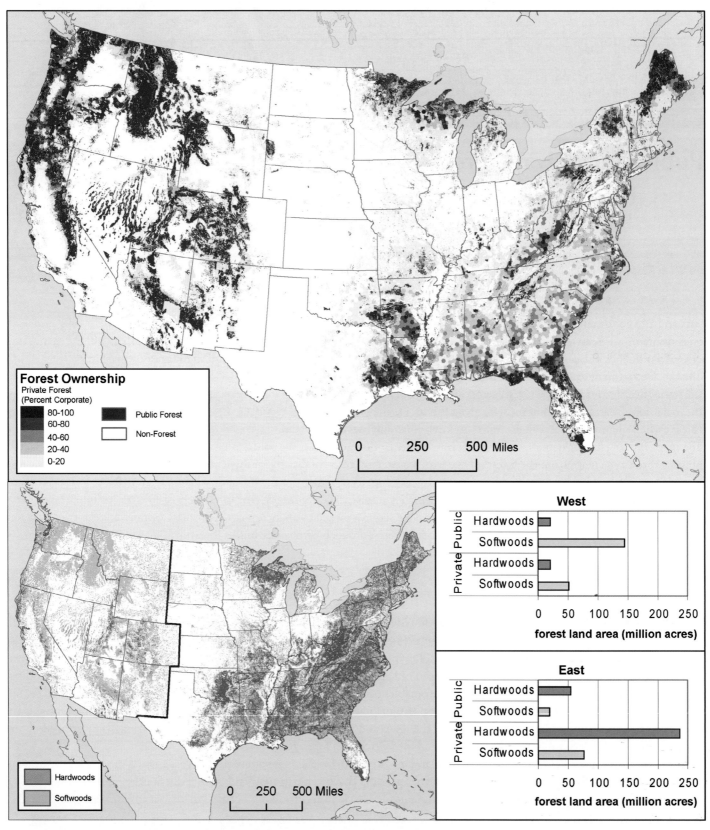

Data courtesy of USDA Forest Service; U.S. EPA; Conservation Biology Institute; ESRI Data & Maps, 2006, from ArcWorld Supplement; ESRI Data & Maps, 2006, from ArcUSA, U.S. Census; ESRI Data & Maps, 2006, from ArcWorld.

Broad patterns are instantly apparent in the upper map, especially the contrast in the amount of private ownership (yellows and reds) in the East as compared to the amount of public ownership (blue) in the West. Also, note the concentrations of corporate private forest (red) which occur in parts of the Northeast, South, and Pacific Northwest.

The bar charts accompanying the lower map were designed so that the colors correspond with the hardwoods (purple) and softwoods (green) map categories. This allows the viewer to quickly associate the bars with the corresponding locations on the map. The bar charts highlight another striking regional difference: the West is composed primarily of softwoods held in public ownership; while the East is composed primarily of hardwoods held in private ownership.

Spatial analyses and maps of forest ownership are powerful tools for rectifying misperceptions about forest ownership, highlighting regional differences, and helping to inform not only public opinion, but public policy, for it is the owners of the forest land who will determine its future use.

Resource tables

Data dictionary

General data description	Data sources
Forest/nonforest softwood/hardwood	Forest Type Groups Map, USDA Forest Service. See the USDA Forest Service Remote Sensing Applications Center Web site.
Percent corporate forest ownership	Resources Planning Act (RPA)
	RPA data on forest ownership is derived mainly from the USDA Forest Service Forest Inventory and Analysis (FIA) program. For more info, see the FIA Web site.
Public land ownership	Conservation Biology Institute's Protected Areas Database, available from the Conservation Biology Institute Web site.
State boundaries, country boundaries, lakes	ESRI Data & Maps.
EMAP hexagons	Environmental Protection Agency. See the EPA Web site.

Software dictionary

Software	Description
ESRI ArcGIS Desktop	Cartographic layout and geoprocessing.
ArcInfo Workstation	GRID module for raster processing.
Microsoft Excel	Bar charts.

Additional resources

Resource	Description and source
Forest Inventory analyst	Able to retrieve, assemble, and process resource inventory data.
GIS specialist	Able to process raster datasets and design the cartographic layout.
ColorBrewer Web site: www.colorbrewer.org	Used to find information on color palettes.

Recipe for map-building success

The following steps outline our thought process and procedures for creating a successful map.

Step 1: Develop the question

Be very specific about what information needs to be communicated. We wanted to answer the question, "Who owns America's forests?" and to rectify the erroneous perceptions about forest ownership.

Step 2: Identify the audience

For these maps, our audience was multitiered: policy makers, concerned citizens, forest landowners, and students.

Step 3: Decide if a map is appropriate

Is there a compelling reason to use a map rather than tables or charts? With regard to ownership, we decided a map quickly gives the user a sense of the amount of forest area in the broad ownership categories of public and private. Furthermore, the regional patterns are quite striking when displayed spatially. We also decided a broad audience would find a map more accessible than a series of tables.

Step 4: Assess data availability

RPA forest ownership data at sample locations was available for the conterminous United States. Using the built-in capabilities of ArcGIS, the forest service researchers were able to quickly transform ownership point data into a usable format and summarize the data on a hexagonal sampling frame, which also maintains landowner privacy. We elected to combine this data with publicly available nationwide datasets on forests (forest-type group) and public ownership (protected areas database from the Conservation Biology Institute).

Step 5: Determine map scale

We evaluated the range of scales/resolutions of the existing data and decided on a common scale.

We wanted to create a map at a continental scale and made a somewhat unconventional choice to intermingle pixel-based datasets with data that was summarized over a unit larger than the pixel. This choice accommodated our need for maintaining ownership privacy.

Step 6: Combine datasets

Datasets were combined via raster processing operations. We used numerous source datasets to create these maps and elected to combine everything into a single raster grid. This greatly simplified work during the cartographic design phase.

Step 7: Design an effective layout

We wanted to communicate several pieces of information: public versus private forest, percent corporate private ownership, hardwood versus softwood, and public hardwood/softwood versus private hardwood/softwood. Furthermore, we wanted to emphasize differences between forests in the East and the West. Early iterations attempted to combine this information in a single map using multiple diverging color ramps. Interpreting the map was difficult, especially when considering color-vision concerns, so we elected to use a pair of maps and complement them with bar charts.

Conclusion

The fusion of ownership data with remotely sensed forest cover data results in an informative, visually appealing, and unique national map that reveals spatial patterns of forest ownership in the United States. By using the Environmental Protection Agency's Environmental Monitoring and Assessment Program's (EMAP) hexagon sampling array we were able to spatially summarize and display plot-based corporate ownership data without disclosing private information about individuals or corporations.

These maps allow the viewer to quickly assess the drastic difference in the ownership patterns between various regions in the United States. The color palettes in the first map were selected to allow the user to easily distinguish between public and private, and the yellow-to-red ramp sharply contrasts private corporate ownership and family/individual ownership.

By including a pair of maps, more information could be communicated while still maintaining simplicity in each map. For example, the forests along the extreme western edge of the Sierra Nevada mountains in California are mostly privately owned (see upper map), and this privately owned area is dominated by hardwoods (see lower map). Such comparisons of the two maps allow viewers to discern how combinations of forest and nonforest cover, public-private ownership, individual-corporate ownership, and hardwood-softwood forests vary geographically.

Analyzing urban forest characteristics in Florence, Alabama

Paul D. Graham, Christopher J. Holder

The city of Florence is located along the banks of the Tennessee River in the northwest corner of Alabama. Considered a primary economic center, Florence is one of the most biologically rich and historically significant communities in the region. To better manage its urban environment and create cost-effective tree maintenance strategies, the City of Florence is using GIS to determine the value and benefit of public trees.

In 2007, city planners conducted their first complete public tree inventory to provide a comprehensive data source for mapping, analysis, and developing management plans that support operational objectives. Operational objectives in Florence include scheduling tree maintenance activities, developing planting strategies to reduce risks associated with invasive species, locating optimum growth sites for soil stabilization, and more. The maps derived from the 2007 tree inventory are used to present findings and discuss development options with elected officials and community organizations.

Collecting data needed to inventory all of the city's trees initially required extensive fieldwork. To create this map, the city collaborated with various stakeholders to pool personnel and technical resources. The team consisted of Bobby Irons, mayor of Florence, and representatives from the U.S. Department of Agriculture (USDA), Forest Service; Auburn University; the Alabama Cooperative Extension System; and the University of North Alabama.

The city hired eleven GIS interns from the University of North Alabama to inventory more than 36,000 trees using Global Positioning System (GPS) devices. Interns measured each tree individually and assessed it for seventeen attributes that describe the tree's physical characteristics, planting site, and condition. Finally, the city's urban forester used the forest service's Street Tree Resource Analysis Tool for Urban Forest Managers (STRATUM) and ArcGIS software to quantify the structure, value, and spatial patterns of the urban forest.

This map and associated graphs provide a clear overview of the combinations of tree species, population, age, and spatial distribution that exist in council district three in Florence. The information helps urban foresters develop strategies to improve age diversity and species distribution. The map is also designed to educate and encourage citizens and local decision makers to participate in urban forestry planning activities.

A visual solution

The map and accompanying graph illustrate the distribution of the city's top ten tree species by diameter class. To determine the age of a standing tree, data collectors often use an increment borer, but using this tool is a time-consuming process. Because most tree species have a documented range of height and crown radius, the trunk diameter can be used to estimate the age of an individual tree using linear regression of the three variables within a species: height, crown radius, and diameter. By focusing only on trunk diameter measurements, the City of Florence collected significant volumes of reliable data for basic and advanced analysis without having to use an increment borer to determine tree age.

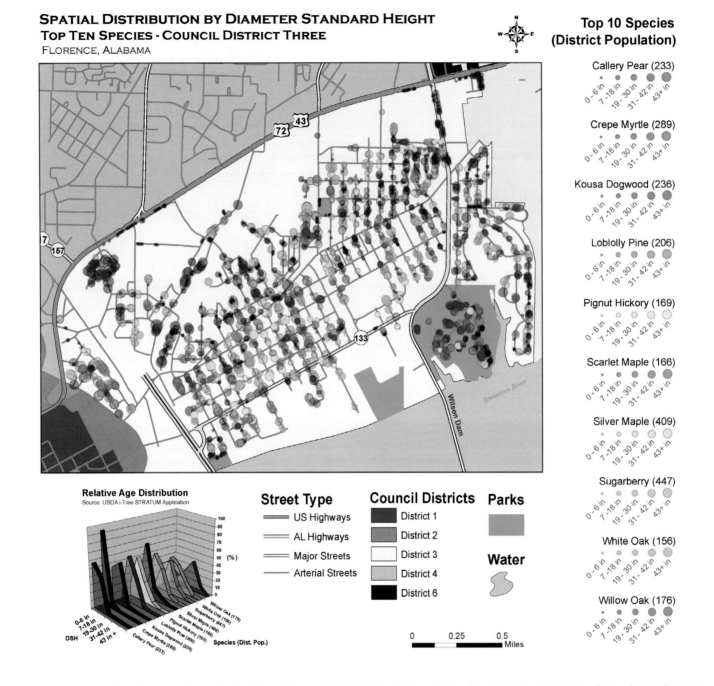

Data courtesy of Paul Graham, Urban Forest, City of Florence; Information Systems Dept., City of Florence; Planning Dept., City of Florence.

Contributors	Organization
Paul Mask	Auburn University, School of Forestry and Wildlife Sciences.
Neil Letson, Stephen McEachron	Alabama Forestry Commission.
Faculty: Bill Strong, Lisa Keys-Mathews, Greg Gaston, Francis Koti	University of North Alabama, Geography Department.
Students: April Blelew, Dustin Durham, Chris Holder, Tina Irons, James Morrow, Randal Reding, Courtney Kelly, Rita Strong, Cathleen Thornton, Kevin White, and Derek Ray	
Paul Graham, Melissa Bailey, Bryan Mitchell	City of Florence, Departments of Urban Forestry, Planning, and Information Technology.

Based on a database summary of the city's tree inventory, district three is home to more than 3,000 individual trees that can be categorized into twenty-eight distinct species. From this summarized database report, the urban forester identified close to 2,500 species into a list of the top ten species within district three. To illustrate these results spatially, the urban forester applied a definition query then used a color ramp to represent the species on the map. A variety of symbol sizes was used to represent diameter.

This map incorporates results generated from the STRATUM analysis model, which allows urban foresters and managers to evaluate the resource structure and economic benefits of trees in an urban setting.

The relative age distribution chart associated with the map summarizes and plots diameter at breast height (DBH) by species. Assuming that DBH indicates age through correlation, the chart best describes the number of trees planted per species over time. The combination of population statistics and spatial traits observed in this map solution suggest:

- a high concentration of large willow oaks in the northwest quadrant
- sugarberry (17 percent) and silver maple (16 percent) comprise a relatively large segment of the overall population (33 percent)
- moving west to east, the population of sugarberries and oaks declines as the population of pines, maples, and crepe myrtle increases

The map uses extensive color ramps to allow the map reader to quickly compare the spatial and temporal data represented, thereby helping city managers to understand where to diversify stand age and species concentrations. Understanding the spatial distribution of species and their respective age within a management area helps to ensure the success of future planting regimes.

As part of a report package, this map is typically accompanied with a brief narrative highlighting key information such as management recommendations, technical and operational positions, and urban forestry development issues.

Resource tables

Data dictionary

General data description	Data sources
2007 Florence public tree survey	City of Florence Urban Forestry Department, USDA Forest Service, Alabama Cooperative Extension System, University of North Alabama.
City limit, parcel, and zoning data	City of Florence, Lauderdale County, Alabama.
Street and highway data	City of Florence, Contractor, G-Squared Photogrammetrists, 20 Ardmore Highway, Fayetteville, Tennessee 37334.

Software dictionary

Software	Description
ESRI ArcGIS Desktop	Spatial and statistical analysis, data display, quality control, and map design.
ArcPad, ArcPad Application Builder	Survey data correction, primarily quality control.
Trimble Pathfinder Office	Primary STRATUM survey creation and data collection, differential correction, data transfer. Provided by University of North Alabama.
USDA Forest Service i-Tree STRATUM	Street tree management and analysis tool for urban forest managers.
Microsoft ActiveSync	Mobile-to-desktop data transfer.
Microsoft Access	Used to house tree survey dataset and additional city data in personal geodatabase.
Microsoft Excel	Used to produce graphs and charts from USDA i-Tree STRATUM analysis and tree survey data.
Microsoft Word	Used to produce i-Tree STRATUM analysis reports and final grant reports.

Additional resources

Resource	Description and source
Trimble Geo-XT GPS units provided by the University of North Alabama	GPS-enabled data logger, collecting waypoints, housing predefined range-based STRATUM survey fields.
Mapstar e-compass and laser rangefinder Laser Technology Inc.	Fully integrated distance and azimuth waypoint correction for GPS receiver.

Recipe for map-building success

This map is intended to be included in official City of Florence report documents. The goal of this project is to produce map solutions that effectively convey operational plans and professional management concepts.

Step 1: Review data
Determine which local spatial datasets are available and investigate whether survey data can be acquired from related agencies, commercial organizations, or community interest groups.

Step 2: Set up STRATUM
Review the data requirements of the forest service STRATUM application. Identify equipment and software needs and determine the type of tree inventory to perform.

Step 3: Identify local support
Identify groups or individuals who can help collect tree inventory data. With limited GIS and GPS staff and technical resources, the City of Florence sought additional capacity through local partnerships.

Step 4: Conduct field survey
Collect field data using GPS to document the location of each tree surveyed. Use coded value domains to define rules about how the data can be stored and edited to ensure data integrity and input validation of survey data.

Step 5: Select management zone to analyze
Determine the management zone to analyze. Perform a definition query using the management zone field and the value of the zone as your parameters. Validate, save, and apply the query. Check that the query is reflected correctly on the map. Open attribute table and summarize results of the species field. Save table to map. Identify top ten species by sorting on count field.

Step 6: Develop tree species query
Perform a definition query to nest the species query inside the saved management zone query.

Step 7: Classify diameter range
Symbolize the tree data subset using the multiple attributes symbolization feature. Apply a color ramp to species name attributes and create a variation in the symbol size by the diameter field. To simplify representation, the Jenks classification method was used in this map with five class break values.

The resulting legend presents a single species with the same color across the DBH class. DBH classes change by a range of symbol sizes. The cartographer converted the legend to graphics so they could better control positioning and spacing of legend elements.

Step 8: Select scale
Select an appropriate data and map scale. For this Diameter Distribution map, the cartographer used a medium scale (1:30,000). This scale provides good spatial orientation and reasonable determination of data attributes.

Step 9: Create a data chart
Create a chart in the layout view based on the STRATUM species summary table for the management area. Modify the species code colors to reflect common names and colors used in the map.

Step 10: Create final map
The final product is a map with supporting information. Legend elements, colors, and symbols should coincide with supporting information. Though this map provides complex information for nonforesters, the map design should be simple and easy to understand for nontechnical people.

Conclusion

Combining numerical and spatial data provides urban foresters with a solid foundation on which to develop a long-range planting plan to improve the mix of tree species. This approach helps reduce the risk of pathogens or invasive species. The planting plan should also improve age distributions and reduce wide spread mortality due to even-aged species populations.

While the map is helpful for understanding the results of the analysis, the numerical information derived from the map makes this a unique product. Including graphs and tables presents comprehensive information on one map that might not otherwise be possible. The portrait orientation of these maps provides better integration into the overall report document and makes viewing them easier for the reader.

For the City of Florence, having an urban forest GIS dataset is a significant benefit. Municipal planners use that data to make decisions and develop and articulate short- and long-term management plans. In short, an urban forestry GIS can be used to display the quantification of monetary and environmental benefits of a healthy and sustainable urban forest.

GIS for Romanian forest management planning

Marius Dumitru, Marius Daniel Nitu, Gheorghe Marin

Romania is in southeastern Europe, bordering the Black Sea and located between Bulgaria and Ukraine. Romania has a long tradition in forestry, having developed its first forest management regulations more than 165 years ago. Forests cover 15.6 million acres, or 27 percent of the total land base. Managed as a public asset, these forests are predominantly found in mountains and hilly regions. The forest is composed of 30 percent conifers, 30 percent beech, 19 percent oak, and the remainder includes a variety of broad-leafed trees.

Romanian forests are divided into forest districts, which are then subdivided into production units. Production units are established based on forest stand composition and similarities related to forest growth, yield, and protection characteristics. For each production unit, the Forest Research and Management Institute (FRMI) or private companies under FRMI's oversight develop a management plan. The FRMI is the government agency responsible for managing, monitoring, and regulating the forest industry.

In Romania, management plans are technical documents that prescribe activities in a production unit to meet economical, social, and ecological objectives. Forest management plans are issued for ten-year durations for the majority of forest types, with plans issued every five years for poplar and willow plantations. Each management plan contains written descriptions of objectives, operational activities, thematic maps, and supporting data. Thematic maps are included to help explain the management plan for each production unit. GIS-based maps are necessary to provide supporting information, and also the technical and activity schedule to implement the five- or ten-year work plan. The maps are also used to verify activities during the plan approval process.

Using GIS in Romanian forest management gives managers the ability to quickly update forest inventory databases and to manage and track activities at any scale required to make strategic, operational, or tactical decisions. Forest management teams are also realizing a significant increase in efficiency and planning accuracy with the use of geospatial technologies.

A visual solution

Forest planning maps represent the cartographic solution used to link descriptive data recorded in a Romanian forest management plan to a precise and accurate location in the field. These GIS maps serve as the operational basis upon which managers and field personnel can identify forest district units and schedule specific on-the-ground activities such as commercial thinning or tree planting.

Maps are initially created at the small-scale forest district level and then, as more data is collected, they are available for larger-scale production units. At the forest district level, GIS maps provide general information about the location of each unit and cadastral features such as ownership parcels and tenure. Additionally, transportation and hydrological features are represented.

The large-scale maps of each production unit, such as the map represented in this chapter, are more detailed than the forest district maps. They include production unit information such as species composition, tree age, production class, consistency, and the activities to complete during the next ten years of the management plan. These maps ensure correct orientation at the subparcel level so field crews can locate their operating areas and conduct their in-forest activities.

Forest Management Planning, Sinaia, Romania

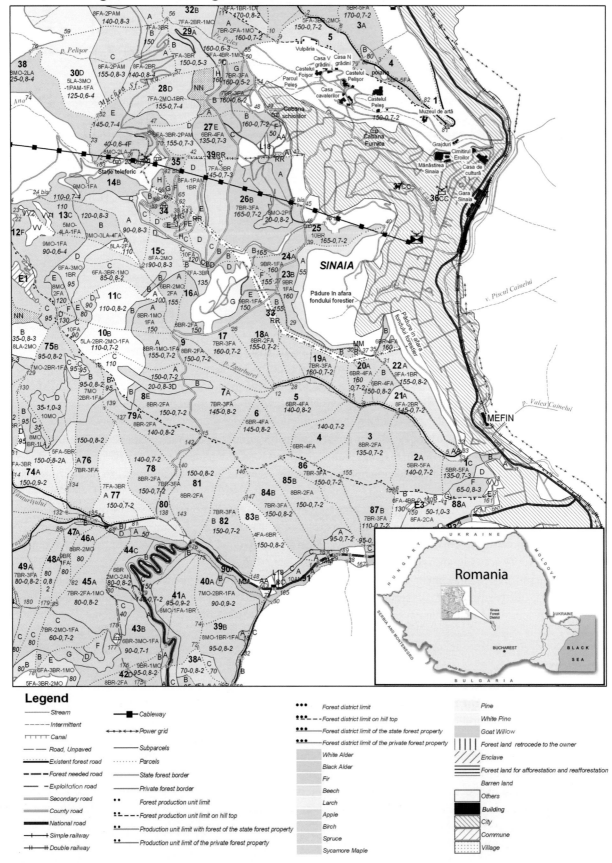

Data courtesy of Marius Dumitru and Marius Daniel Nitu.

Forests are prone to frequent and intense changes such as those related to disturbances such as harvesting or wildfires. The use of GIS enables land managers and foresters to quickly update maps as changes occur in the field. The affected maps can also be efficiently reproduced at various scales in an ad-hoc manner as required. This approach reduces the time required to develop new management plans as inventory is continuously updated instead of having to re-create a new map every time plans are developed or new strategies are being considered.

The development and maintenance of a GIS database provides numerous opportunities for additional forestry management planning scenarios outside of the classical inventory approach. Such revenue generating opportunities can be realized by using GIS to estimate distances from or to timber stacks, road maintenance, calculating annual yield, and more.

Resource tables

Data dictionary

General data description	Data sources
Topography maps 1:5,000 equipped with forest planning elements	Forest Research and Management Institute Bucuresti (ICAS).
Topography maps 1:10,000 equipped with forest planning elements	Forest Research and Management Institute Bucuresti (ICAS).
Topography maps 1:25,000	Military Topographic Department (DTM).
Orthophotomaps	National Geodesic Institute.

Software dictionary

Software	Description
ESRI ArcGIS Desktop	Basic functions with topology and advanced editing.
ArcGIS Publisher	Forest districts from National Forest Administration—RNP-ROMSILVA.
Microsoft Access	Forestry database management system
ERDAS IMAGINE	3D model building and analysis
GIS manager	Engineer specializing in GIS (trained by ESRI Romania) and familiar with database and query techniques.
GIS specialist	Engineer or technician specializing in GIS (trained by ESRI Romania) and familiar with digitizing techniques.

Additional resources

Resource	Description and source
Descriptive database from forest planning software (special application called AS)	Provide special database for GIS Department—Forest Research and Management Institute Bucuresti (ICAS).
Topographic maps	Special maps with forest limits, stone marks, and forest roads—Forest Research and Management Institute Bucuresti (ICAS).

Recipe for map-building success

Step 1: Develop cartographic standards

The Forest Research and Management Institute (FRMI) developed cartographic standards for GIS maps that are similar to the institute's original hard-copy maps. Colors, line types, and symbols were defined to ensure that all users would have access and be able to use the new standard. Additionally, visual characteristics were considered such as the order of layers, layer transparency, the naming convention for each layer, and label displays.

Step 2: Develop geodatabase model

FRMI GIS specialists developed the geodatabase model to reflect the needs of forest planning activities.

Step 3: Assemble database

Historical base data in analog format was scanned so it could be converted to a digital environment. Hard-copy maps scanned and converted included topographic maps, legal parcels, reference markers, aerial photographs, orthophotos, and so on. Each map was georeferenced into the Romanian national projection system (Stereographic 1970).

Step 4: Vectorize hard-copy maps

Cartographic elements only available in hard-copy format were converted for use within a GIS using heads-up digitizing and on-screen vectorization. After raster vectorization, each element was attributed according to each feature type.

General cartographic elements that are depicted as vector lines include hydrography, communications infrastructure, roads, markers, and more. Elements that are best represented by polygons were also vectorized. These polygon features include forest boundaries, building footprints, forest cover, and more.

Step 5: Add attributes

Each new feature added to the GIS required specific coding to record the attributes of each point, line, or polygon then labels were added.

Step 6: Distribute data to users

Data was converted using ArcGIS Publisher and shared among staff in forest district offices throughout the Forest Research and Management Institute. Personnel were trained to use ArcReader to view and print data.

Step 7: Create new data

New maps were then created for further analysis using existing data. Maps to show slope, aspect, erosion, and solar radiation were created based on the attributes of each source layer.

Conclusion

GIS in Romanian forestry planning allows for significant benefits at both the strategic planning and operational levels. Newly compiled forestry planning maps are more understandable than those developed using traditional cartographic drafting methods. The ability to print these maps at different scales and at any time they are needed is a significant incentive for the Forest Research and Management Institute to invest in geospatial technologies.

The forest stand map in this chapter is considered one the most important maps in Romanian forestry planning because it clearly describes the forest inventory for each production unit. Color and labels represent complex information about each forest stand in an easy-to-understand format. For instance, the tree species is represented by color codes, while tree composition, age, production class, and consistency are represented by labels.

With the development of a GIS for FRMI's management planning department, numerous new options for analysis can now be realized, in part because of the creation of new datasets, but also because the software offers tools for additional investigation and exploration. GIS is now being used for managing forestry planning, protection, leases and encumbrances, to name a few.

Due to the relationship between the map and database environment, all labels are derived automatically without having to reprint or manually insert each label. Prior to using GIS, this was a time-consuming task that led to many versions of the same map. With the development of the forestry GIS, updates can be completed as they occur, providing managers with a current dataset upon which to base their decisions.

Using a topographic index to define terrain types

Heinrich Goetz

Stretching from southern Kansas into central Texas, Cross Timbers is a semiarid transition zone between the prairies of the Great Plains and the forested Southeast. Cross Timbers is a mix of oak woodlands and savannahs, tallgrass prairies, and bottomland hardwood forests. Post and blackjack oak dominate the woodlands and savannas, and the bottomland forests contain a mix of ash, elms, sugarberry, pecan, and other hardwood species. The primary drivers of this vegetation pattern are soil texture and hydrology. Oak woodlands and savannas occur on coarse textured upland soils, tallgrass prairies occur on fine textured upland soils, and tree species composition in the bottomlands varies with soil texture.

Understanding flooding patterns and soil moisture dynamics within a region such as Cross Timbers is essential to protecting water quality and to the successful restoration and creation of riparian ecosystems. The map in this chapter was created to predict vegetation terrain types in an area where species composition varies with hydrology and soil texture.

This map helps land managers and researchers visualize and understand soil moisture dynamics. It is used in ecological restoration efforts to determine the most appropriate areas for planting species with various moisture requirements or anaerobic tolerances and also to identify and map candidate locations for wetland restoration. As the basis for further analysis, this map was applied in a forest gap model for a watershed within the Cross Timbers ecoregion to simulate forest dynamics by modeling tree growth on a plot-level scale. Model input parameters included tree species characteristics, weather, topography, and soil.

A visual solution

This map provides a method for GIS analysts and researchers to partition watersheds into terrain type units that reflect different soil and hydrological conditions. By delineating terrain types across a landscape and combining soil type with a measure of topographic position, GIS models can be used to predict forest and vegetation patterns.

Topographic position is represented on this map with a compound topographic index (CTI) layer. This layer is generated from a 3D digital elevation model (DEM) using ArcGIS Spatial Analyst tools. The CTI is also known as a topographic wetness index that predicts moisture accumulation in resulting areas that are either prone to flooding or where the soil contains a high moisture content.

Analysts used the Terrain Analysis Using Digital Elevation Models (TauDEM) extension developed by David Tarboton at Utah State University to calculate flow direction using its D-infinity algorithm. Flow accumulation and specific catchment area layers are generated from the flow direction, and slope is generated from the DEM.

This map provides a method of dividing the watershed into terrain type units, where each terrain type has a unique set of gap model parameters. The hydrological effect of topographic position of a plot within the watershed is simulated using a moisture run-on coefficient, which is estimated using the CTI values in that terrain type. The map provides a method of extending the forest gap model to simulate the vegetation patterns resulting from topography and soil type.

This map is successful because it provides a visualization of soil-topography-water interactions. It also provides a method for using GIS layers to delineate terrain types. The design highlights the importance of the combination of soil and topography on conditions for vegetation.

Using a topographic index to define terrain types

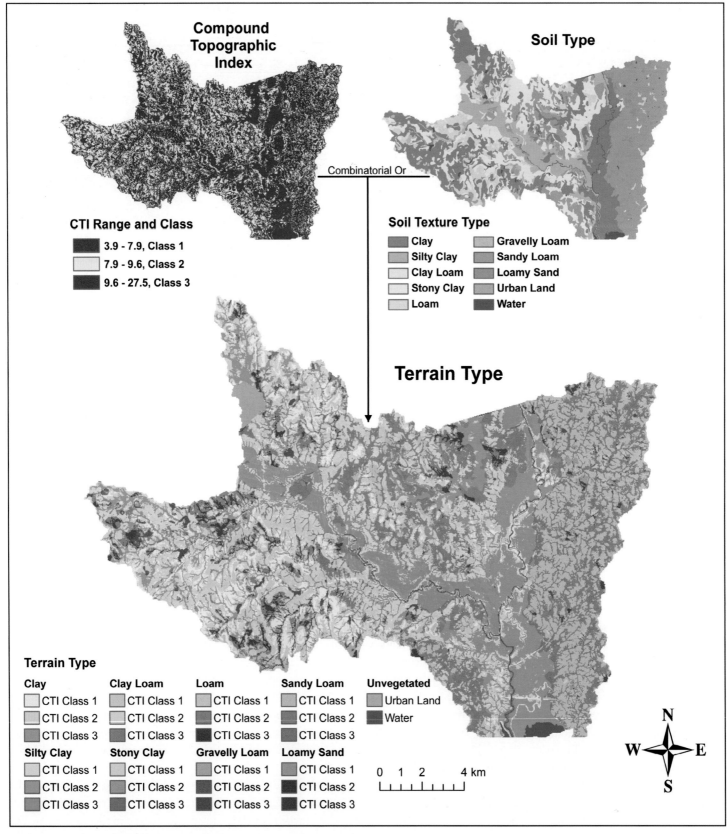

Data courtesy of U.S. Geological Survey, EROS Data Center; Heinrich Goetz; Soil Survey Staff, USDA Natural Resources Conservation Service; Soil Survey of Denton County, Texas [accessed January 29, 2008].

Resource tables

Data dictionary

General data description	Data sources
1/3 arc second (~10 m) digital elevation model	U.S. Geological Survey National Elevation Dataset from the National Map Seamless Server.
Soil survey map and tables for Denton County, Texas	USDA Natural Resources Conservation Service Soil Data Mart.

Software dictionary

Software	Description
ESRI ArcGIS Desktop	Used for determining locations and parameters of model terrain types.; map display and overlay of data layers and aerial photographs.
ArcGIS Spatial Analyst extension	Raster calculator, reclassify, conditional and combinatorial functions; conversion tools for feature to raster and raster to ASCII.
Microsoft Access	Needed to access the USDA Soil Survey tabular data.
Terrain Analysis Using Digital Elevation Models (TauDEM) extension for ArcGIS	Used for its D-infinity method of determining Flow Direction.

Additional resources

Resource	Description and source
Technical article, "Modeling forest landscapes: Parameter estimation from gap models over heterogeneous terrain."	Acevedo, M. F., S. Pamarti, M. Ablan, D. L. Urban, and A. Mikler. 2001. Modeling forest landscapes: Parameter estimation from gap models over heterogeneous terrain. Simulation 77(1-2):53-68.
National Agriculture Imagery Program (NAIP) aerial imagery	Used to overlay terrain types and model results with existing vegetation cover; courtesy of the Texas Natural Resources Information System.

Recipe for map-building success

This map, used to generate forest gap model terrain types across a landscape, is created by combining soil types with categories of a CTI.

Step 1: Delineate watershed boundaries

Obtain a DEM for the region of study and delineate watershed boundaries using the DEM as the source layer. The landscape boundary in this map was derived from a DEM obtained from the U.S. Geological Survey (USGS).

Step 2: Rasterize soil texture map

Obtain a soil type layer and convert the soil map units into a raster layer. The soil layer in this map was sourced from the USDA as a shapefile.

Step 3: Generate CTI classes

Generate raster slope, flow accumulation, and CTI layers from the source DEM. Review the distribution of values of each within each of the soil types to determine an appropriate number of CTI categories. Using the Raster Calculator, generate the CTI. Use quantiles to group the CTI into three classes.

Step 4: Generate terrain type

Combine the CTI classes with the soil type layer using Spatial Analyst's combinatorial or function, and define each combination of soil type and CTI class as a terrain type.

Step 5: Generate run-on parameter

Through analysis of the distribution of flow accumulation and slope values within each terrain type, determine an appropriate algorithm for generating the run-on parameter.

Step 6: Analyze results

Analyze model results by mapping output for each terrain type and comparing with vegetation surveys taken in the study area and mapped with GPS coordinates, or by comparing with aerial photographs or other remotely sensed image.

Step 7: Develop map

Create a map suitable for printing. Include legends to describe the terrain types.

Conclusion

The elements of this map are based on two physical factors affecting forest growth: soil conditions and the movement of water across the surface. These are accounted for by combining soil types with CTI classes to define terrain types.

Analysts combined soil texture and a compound topographic index to create the terrain type map because in a dry environment such as the Cross Timbers, the movement and accumulation of water and the soils' ability to store moisture are crucial to the types of vegetation that can be supported. Soil texture was chosen for its role in water storage capacity and hydraulic conductivity. The CTI is used as a measure of soil moisture accumulation because flow accumulation alone does not identify moisture that accumulates on bottomland flats outside of stream channels.

In some locations, such as in mountainous areas, other ecologically relevant factors such as elevation and aspect must be considered in defining terrain types. Using a USDA soil map allows researchers to use published soil surveys to estimate soil parameters for the model.

Using the CTI rather than both slope and flow accumulation also reduces the total number of terrain types. The strength of this map is its algorithmic method for predicting vegetation terrain types and partitioning a landscape into smaller units suitable for setting parameters in a fine-scale forest model. The design of the map illustrates the simple idea of combining two layers, CTI class and soil type, to yield insight into the complex interactions of topography, soil, and water.

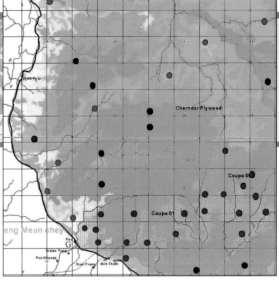

Analyzing the forest structure in Northern Cambodia

Dana Kao, Tsuyoshi Kajisa, Nobuya Mizoue, Shigejiro Yoshida

The northern Cambodian forest in the province of Preah Vihear is located northwest of the historic temples at Angkor Wat and the tourist city of Siem Reap. Bordering Thailand and Laos, Preah Vihear is one of the most remote and sparsely populated provinces in Cambodia with significant areas of natural deciduous dipterocarp forests. The mainly tropical lowland rain forest trees reach 130 to 230 feet tall and are important to the logging industry throughout Asia, South America, and Africa. The province also has seasonal wetlands, large grasslands that flood during the rainy season, and dense evergreen forests found along watercourses and in upland regions.

Since early 1996, Cambodian forests have been cleared, temporarily developed, and then abandoned for shifting cultivation agricultural use. In 1999, the government attempted to better manage the forest in Preah Vihear, but was unsuccessful because of a lack of forest inventory and forest stand data. Without this information, managers were unable to make decisions based on the principles of sustainable forest management.

The Laboratory of Forest Management at Kyushu University in Fukuoka, Japan, was asked to create this map to help forest managers make decisions. Originating from satellite imagery and then field verified, this map is a complete inventory of tree density structures and forest type classification in Preah Vihear province. Using the map, managers have better and more accurate information upon which to base their plans. Additionally, this map shows the geographical location of sample plots and other infrastructure.

A visual solution

The primary objective of this map is to help local stakeholders predict the productive capacity of the Preah Vihear provincial forests. Research scientists at Kyushu University used satellite images to identify forest inventory polygons generated from preliminary unsupervised classification. Initially, deciduous, logged evergreen, mixed evergreen, and dense evergreen forestland was identified.

Planning for a sustainable forest industry meant that a baseline forest inventory was needed. Scientists used image classification to develop the first version of a comprehensive forest inventory. To reduce errors inherent in remote sensing classification and interpretation, ground verification and statistically valid field surveys were completed to refine the initial inventory. Forest technicians identified representative locations of potential permanent sample plots for field-based data collection. In the field, the locations of each of the primary sampling units were found by means of a Global Positioning System (GPS). The project team established 540 field-based inventory sample plots on 80 primary sampling units to verify forest cover classified by remote sensing analysis.

In the field, technicians conducted numerous surveys to identify both forest and nonforest attributes. Forestry surveys included biotic observations and measurements related to tree species identification, height, diameter, age, and canopy closure. Abiotic observations included terrain, elevation, soil type, ground condition, transportation corridors, and so on. The field team also identified areas that had been illegally logged as well as nontimber values such as wildlife and indigenous use.

Analyzing the forest structure in northern Cambodia

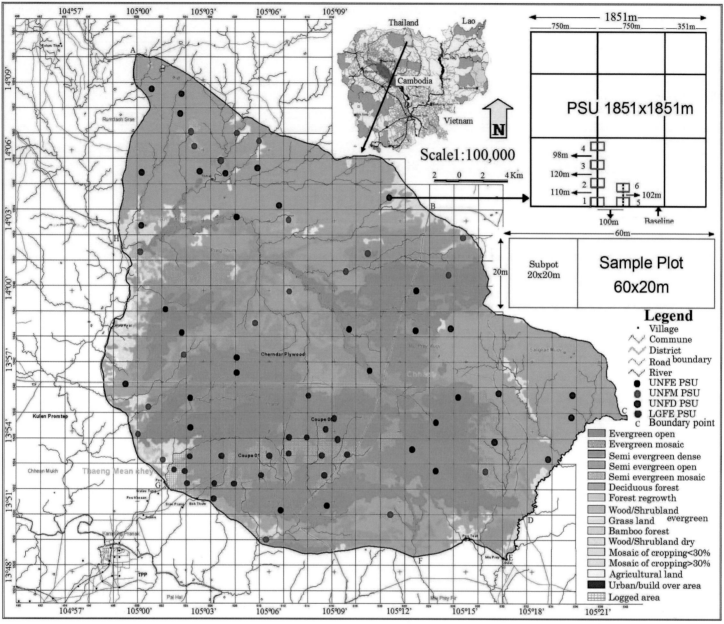

Data courtesy of Dana Kao.

Once the forest inventory map was updated to reflect on-the-ground measurements, important GIS layers were added, including land-use classification, land-use boundaries, and forest structures. Managers now use this map as a guide to verify and assess ground vegetation for updating, planning, and zoning the forest, based on density or volume. This map is also critical for comparing the error between satellite imagery and ground vegetation, which improves the accuracy of the provincial forest inventory.

Resource tables

Data dictionary

General data description	Data sources
2000-2001 forest inventory	Cambodia Cherndar Plywood Company.
Land ownership and concession boundary	Cambodian Forestry Administration.
Forest cover from satellite image 1997	Cambodian Forestry Administration.
Forest cover from satellite image 2000	Laboratory of forest management, Kyushu University. Compare the imagery of years 2000 and 1997.
Primary sampling unit (PSU) locations	Cambodia Cherndar Plywood Company.
Plot locations	Cambodia Cherndar Plywood Company.
Infrastructure: roads, boundaries, and rivers	Images from Laboratory of Forest Management, Kyushu University and Cambodian Forestry Administration.
Land use	Cambodian Forestry Administration.
Geographical data	Cambodian Forestry Administration.
Disturbed forest area	Cambodian Forestry Administration.

Software dictionary

Software	Description
ESRI ArcGIS Desktop	Spatial analysis tools.
Microsoft Excel	General data analysis from the field.

Recipe for map-building success

Step 1: Data procurement

Satellite data for the province of Preah Vihear was obtained, orthorectified, and projected into the universal transverse Mercator (UTM) coordinate system. Base GIS layers such as hydrography, roads, and administrative boundaries were included to aid in field navigation.

Step 2: Satellite data processing

Using existing land-cover maps, satellite images across the study area were processed to determine forest cover changes caused by human activity or natural occurrences.

The newly obtained data was then processed automatically using the unsupervised classification technique to determine land-use classifications. Because unsupervised classification results in more categories than required, field surveys were planned to determine appropriate groupings to develop an operational forest inventory map.

Step 3: Field sampling planning

One hundred primary sampling units (PSUs) were placed on the map for each forest type. Twenty PSUs were then randomly selected in each forest types to represent sample plots. The universal transverse Mercator (UTM) grid coordinates were noted for field crews to locate the plot center of each sample plot.

In total, 520 sample plots were randomly selected and included coverage in evergreen forests (160), mixed evergreen (140), deciduous (100), and logged evergreen (120).

Step 4: Field data collection

Field technicians found the location of each plot center using a GPS device. They created 20m x 60m plot boundaries and measured and assessed the trees in each plot, collecting attributes such as species, stem form, diameter, height, and more on trees with a minimum diameter at breast height (DBH) of 10 cm.

Step 5: Data loading and verification

Analysts updated the datasets based on data collected in the field in each sample plot. In ArcGIS, the map was updated to reflect the forest cover type and density observed on the ground. Additionally, land-cover changes that occurred between the time the satellite image was taken and when data was collected in the field were noted and the respective polygons were also updated.

Step 6: Weighting factors

Analysts applied statistical mathematics to estimate the density structures by forest type so they could better understand the species mix found in each plot. Each forest type was then classified into a variety of different polygons. Attribute data values were loaded into the geodatabase and related to each polygon, including density structure, forest type, and land use. For each forest cover polygon, ArcGIS calculated volumes in cubic meters and area in square meters.

Step 7: Zoning delineation

A zoning management code was automatically generated by first querying each forest cover polygon based on area and volume, then selecting and attributing polygons that met a predefined look-up table. Polygons were then converted into a raster layer for further geospatial analysis.

Step 8: Data smoothing

To reduce the error of overlapping forest cover and agricultural land, analysts applied a moving average filter to smooth errors that occurred. Based on stakeholder consultation, a map was created with the new administrative boundaries.

Step 9: Symbolization and mapping

A final map was created for stakeholders to use in their forest planning activities.

Conclusion

This map was initially used to plan field data collection activities and eventually it evolved into an accurate forest cover map for the province of Preah Vihear. Originally based on unsupervised classification of remotely sensed data, the derived forest cover was then verified and modified based on extensive statistical sampling methods.

The forest inventory map, used in combination with administrative boundaries, infrastructure, villages, and other base mapping information, provides information necessary for strategic planning activities. Additionally, stakeholders now use this map to estimate forest cover by type, classify watersheds, identify the location of provincial infrastructure, and monitor canopy change over time. At the operational level, this map also serves as an accurate reference for field crews working in the area.

This project was initiated to predict the productive capacity of the province's forest structure. The final map presented in this chapter is a significant step in providing forest managers with a tool to support decision making for both harvesting and protection. Additionally, this map will help local communities better understand where activities will be taking place and the impact that land cover change will have on their forests.

Using an integrated moisture index to assess forest composition and productivity

Matthew Peters, Louis R. Iverson, Anantha M. Prasad

The 834,000-acre Wayne National Forest, Ohio's only national forest, lies in the rolling foothills of the Appalachians in the state's southeast. Congress established the forest boundary in 1934 to prioritize land acquisition and ownership of forest lands in need of restoration. The forest is composed of both central hardwoods, primarily oak and hickory, and softwoods, including native pine and hemlock.

To determine long-term soil moisture conditions, a portion of the Wayne National Forest was analyzed to create an integrated moisture index (IMI) map. The IMI was then used to predict forest site productivity and composition. IMI maps are a useful tool for managers and researchers to stratify treatment units so they can understand the effects of prescribed fire and commercial thinning in southern Ohio. When combined with forest inventory and analysis (FIA) data, IMI can provide additional information on site quality and the basal area of various tree species.

The IMI values ranging from wet (green) to dry (brown) are influenced by four layers, including topography, slope, the movement of water, and water supply availability. Using GIS, these layers can be combined and analyzed to predict the forest's productivity and species composition. Topographic maps provide information about the form and elevation of the landscape. Slope maps provide information about the steepness of a landscape, either as a percentage or in degrees. Mapping the movement of water helps to identify both the source and destination of moving water. Finally, available water supply shows the amount of moisture that can be stored in the soil.

The map of IMI is a result of a weighted calculation of the four layers, where each map layer is weighted according to its influence on long-term soil moisture. Flood boundaries are used to mask locations where IMI produces unreliable values because it is difficult for a GIS to correctly determine the direction of water movement across a flat landscape.

A visual solution

The integrated moisture index analysis can predict a forest's site composition when species-specific site conditions are related to index values. One would expect to find species known to perform well on dry sites to be present where IMI values are low and species that prefer moist sites where values are high. With this information, strategic plans can be developed to conserve individual species or groups of species that require specific site conditions. Mapping IMI also gives forest managers a visual assessment of the flow and storage of water across the landscape in the form of a map.

Developing an integrated moisture index map requires extensive use of raster-based analysis. Raster data consists of rows and columns of cells. Each cell stores a single value such as percent slope, aspect, elevation, direction of flow, to name a few. Cells are the smallest individual unit of a raster layer, and the resolution of a raster layer is dependent on the ground scale that a single cell measures, usually calculated in meters. GIS analysts performed raster analysis using the ArcGIS Spatial Analyst extension to generate the IMI model. Specific GIS functions used include hillshade, curvature, and flow accumulation.

Researchers also used IMI to stratify the landscape of four treatment units in a study of oak regeneration using prescribed fire and mechanical thinning. Locations where IMI indicated dry soils had more intense fires, greater canopy openness, and increased oak and hickory regeneration over a seven-year period.

Integrated Moisture Index: Wayne National Forest

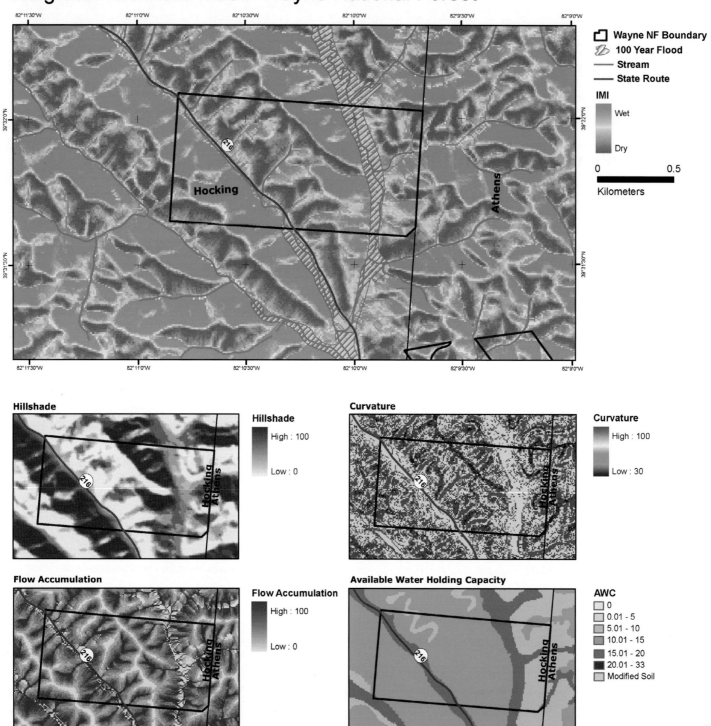

Data courtesy of Ohio Department of Natural Resources Division of Geological Society; U.S. Department of Agriculture, Natural Resources Conservation Service; Ohio Department of Natural Resources Division of Water; U.S. Department of Transportation, Federal Highway Administration; Ohio Department of Natural Resources Division of Real Estate and Land Management; USDA Forest Service, Wayne National Forest.

Resource tables

Data dictionary

General data description	Data sources
10m DEM of Ohio	Ohio Department of Natural Resources.
County soil surveys	USDA Natural Resource Conservation Service.
Wayne National Forest polygons	Ohio Department of Natural Resources.
FEMA 100-year flood polygons	Ohio Department of Natural Resources.
State routes	Ohio Department of Transportation.
Streams	Ohio Department of Natural Resources.

Software dictionary

Software	Description
ESRI ArcGIS Desktop	Cartographic production, data management and analysis.
ArcGIS Spatial Analyst extension	Raster analysis.
ArcGIS Maplex extension	Provides advanced label placement for ArcView.
Soil Data Viewer 5.2	ArcGIS extension that maps soil attributes.
TauDEM	ArcGIS extension that provides advanced analysis tools for processing a digital elevation model (DEM).
Hawth's Tools	ArcGIS extension that provides various tools for ecological research.

Recipe for map-building success

The integrated moisture index can be created from a digital elevation model (DEM) and county soil survey data. A DEM represents the surface topography of the earth to help define both the form and elevation of the terrain. The quality of the output is determined by the resolution of the DEM and algorithm used to calculate flow accumulation. Our model uses TauDEM's D-infinity algorithm to calculate flow direction and flow accumulation from a ten-meter DEM. We used the USDA Natural Resources Conservation Service's Soil Data Viewer to calculate the soil component of the IMI model. Once the area had been selected and the data acquired, we followed these steps:

Step 1: Apply the Fill tool to reduce errors
From the Spatial Analyst toolbox, we ran the Fill tool on the DEM to reduce errors in the model input files.

Step 2: Calculate hillshade
We calculated hillshade and curvature from the filled DEM using tools in the Spatial Analyst toolbox.

Step 3: Standardize hillshade output
We used the raster calculator to standardize the output of hillshade and curvature from 0 to 100.

Step 4: Calculate flow direction and slope
We used TauDEM to calculate D-infinity flow direction and slope. The output was in radians.

Step 5: Determine contributing area
With the flow direction, we used TauDEM to calculate the D-infinity contributing area with a slope weighting.

Step 6: Calculate flow accumulation
We standardized the output of contributing area (flow accumulation) from 0 to 100 using the raster calculator.

Step 7: Map available water supply
Using the Soil Data Viewer extension, we mapped the total available water supply, 0 to 100 cm. We then converted the available water supply vector file to a raster grid.

Step 8: Calculate integrated moisture index
We used the raster calculator to calculate IMI with the following formula: [(hillshade x 0.4) + (curvature x 0.1) + (flow accumulation x 0.3) + (available water supply x 0.2)].

Step 9: Mask flat areas
We overlaid flood data to mask areas of unreliable values due to flat topography.

Step 10: Label features
We labeled features using the Maplex extension, which provides increased label control over placements.

Conclusion

Stratifying a landscape with the integrated moisture index can help with sample design and provide information related to soil moisture. The IMI is algorithmically simple to produce, requires no field data, and is reliable for areas with moderate topography. It is not appropriate for flat or mountainous areas where elevation is a prime driver of vegetation. The resulting model is not time-specific and is consistent among locations. Using the D-infinity algorithm produces a more representative pattern of water movement through the landscape. The model can be calibrated to the landscape influences specific to your area by modifying the IMI formula weights.

CHAPTER 12

Mapping land use at Swanton Pacific Ranch

David Yun

Swanton Pacific Ranch (SPR) is a 3,200-acre educational and research facility in the coastal mountains north of Santa Cruz, California. The ranch contains a number of different landscapes, including forests, range, agricultural crops, wetlands, and developed areas with structures.

The ranch was donated by the late Al Smith, successful businessman and California Polytechnic (Cal Poly) alumni, who wanted the property to be maintained as a working ranch and used exclusively for agriculture, recreational, and educational purposes. California Polytechnic State University, San Luis Obispo, College of Agriculture, Food, and Environmental Sciences manages SPR like a commercial ranching operation with crop, grazing, and forestry programs. The ranch also provides a unique "learn by doing" environment to teach sustainable resource management. Faculty and graduate students actively pursue research opportunities in the ranch's forest, range, and watershed areas.

The various land uses, including educational, recreational, and commercial operations, draw numerous people to the ranch. Many interns stay for several months and students visit regularly for assorted class projects. Graduate students and faculty members perform multiyear research projects at the ranch. Recreationists enjoy hiking, horseback riding, and mountain biking in the ranch's varied topography and land cover. The ranch is also used as a meeting and demonstration location for visitors from other educational institutions or governmental agencies. With so many visitors, an accurate map showing all administrative boundaries in detail with reference information is necessary.

The natural resource department's GIS specialist created this basemap; it is used to assist ranch managers to clearly communicate with project managers, stakeholders, and the general public. It is much easier for ranch managers to identify features on the map than try to explain the location of research sites, forestry types, and more. This basemap is also used to depict research projects and forest management activities that are conducted on the ranch. An example of this is the overlay of the Little Creek Watershed project shown on top of the land-use map.

An orthorectified aerial image was purchased in 2005 to identify and create GIS layers for land-surface features. The orthophotos are now extensively used as background imagery for many projects at the ranch. Lidar (light detection and ranging) data, originally purchased for watershed analysis, was used to create the elevation model and the hillshade raster. Although it is used to show the terrain in this map, lidar data is also being used to detect many land-surface features such as forest roads that are hidden by the trees, landslide deposits, and hazardous slopes.

A visual solution

The entire Swanton Pacific Ranch is divided into five major land-use types that correspond to different landscapes. The land supports three main ranch operations: forestry, livestock, and crops. The forested lands are predominantly second-growth redwood and Douglas fir forests with some Monterey pines on the coastal side of the forest. Most trees were heavily logged in the early 1900s when San Francisco required timber to rebuild after the devastating earthquake and fire. Regular timber harvesting activities continued to the 1970s.

Land Use of Swanton Pacific Ranch
California Polytechnic State University

Legend

- ⊕ CFI Points
- ▬ Highway 1
- ── Paved Road
- ---- Dirt Road
- ┼─┼ Railroad
- ── Perennial Stream
- ---- Intermittent Stream
- ☐ Crops
- ☐ Forest
- ☐ Natural Areas
- ☐ Range
- ☐ Structure Areas
- ☐ Management Units
- ☐ Parcels

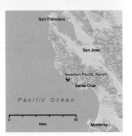

Data courtesy of David Yun; ESRI, developed using GTOPO30, Shuttle Radar Topography Mission (SRTM), and National Elevation Data (NED) data from the U.S. Geological Survey. Copyright 2008 ESRI.

Forestry is still the predominant commercial operation at Swanton Pacific Ranch. Livestock operations occur in the range land located on the western side that overlooks the Pacific Ocean. Typical livestock operations are stocker, cow-calf, and sheep. Cropland, located in the southern part of the ranch, produces organic vegetables. Wetlands and other natural areas provide excellent educational tours for students and visitors. Structural areas consist of houses, barns, sheds, and any other built structures. All five major land-use types are further divided into management units that are determined by their location and usage. For example, the range land-use type consists of several holding areas and pastures by their location. SPR comprises a total of seventy-five management units.

Continuous forest inventory (CFI) points are used as referenced locations for many land management activities such as mapping forest lands, selecting sampling trees, and performing timber inventory. These points are also located on the ground as rebars with a yellow cap and metal identification tags, placed in a 500-foot-square grid pattern. On the map, a parcel layer provides legal property boundaries on the ranch, identified with an assessor's parcel number (APN) to clearly mark the locations of each parcel.

With the exception of an overview map, created using ArcGIS Online Services, the aerial imagery, and the lidar data, all other data was created by Cal Poly State University students. They created the data as part of a senior class project held on the ranch or in an advanced GIS class. The ranch managers and GIS specialist performed quality control and quality assurance tests before using this data as part of the base GIS layer. In addition to providing features, mapmakers annotated several well-known geographic locations for quick reference.

The map was designed for ANSI C (17 x 22 inches) and can be used as either a wall map or a working field map. The wall map provides an excellent way to communicate with visitors about project locations. The map is also used as a basemap that users can quickly modify to showcase their specific projects. Overall, this is a multipurpose map that can be used to clearly communicate land-use management issues and activities.

Resource tables

Data dictionary

General data description	Data sources
2005 aerial image	Digital Globe.
Parcels	Digitized using assessor's map.
Roads	Digitized using aerial and lidar data. Local government may have many roads mapped; however, many ranch roads are usually not mapped.
Streams	Digitized using aerial and lidar data.
Land use	Created by ranch managers. This is Swanton Pacific Ranch-specific land-use data.
Continuous forest inventory	Created from survey information.
ESRI_ShadedRelief_World-2D	ArcGIS Online resources.
Lidar data	Acquired for watershed research project.

Software dictionary

Software	Description
ESRI ArcGIS Desktop	To create and update all layers and create the map.
ArcGIS Spatial Analyst extension	To create the elevation grid and the hillshade raster layer.

Additional resources

Resource	Description and source
Trimble 4800 GPS	To collect survey data for establishing continuous forest inventory points.

Recipe for map-building success

To create a similar land-use map, follow these steps:

Step 1: Determine layers needed

Conduct a meeting with ranch managers to identify which features to put on the map. Since putting every available layer on a map is not feasible, determine the set of desired layers for the map.

Step 2: Obtain layers

Determine available layers and what additional layers are needed. County and local governments may have data layers that are needed for the project. Obtain all necessary layers. In our case, many students created data and a GIS specialist checked it for spatial and attribute accuracy.

Step 3: Edit layers

Edit any layers that need adjustment. Spatial accuracy requirements are different for layers that are produced by other agencies. Update all layers to your accuracy standard.

Step 4: Present layers

Organize layers for display. Display land-use and parcel layers as hollow fill to allow the aerial image to show through. Symbolize other point and line features so that they can be seen clearly and distinctively. Experiment with several colors; it is difficult to pick a color that will work well in all areas since aerial image is used as background.

To achieve a better cartographic effect of showing the boundaries with semi-transparent border, we created a separate land-use type layer. Applying the center as hollow and making thick boundaries for the land-use types allows much of the background information to show through while clearly delineating boundaries. While many professional cartographers use Adobe Illustrator to create visually pleasing maps, we decided to use only ArcGIS Desktop for mapmaking so that many students trained in GIS could update and modify the map. Here are some additional tips:

- Use transparency to show multiple layers. In this case, the aerials are visible under the thick land-use type layer boundary.
- Create a balanced look by placing the map elements using align and distribute tools.
- Use scale units and divisions that make sense. Set the scale bar division values to rounded numbers. For example, if a default scale could be divided into 1,105 meters, change to 1,000 meters for a cleaner look.

Conclusion

Ranch managers at Swanton Pacific Ranch need an accurate land map that has enough detail for operational use. With so many people visiting the ranch, identifying locations within the ranch is a regular requirement. The map clearly shows all possible land use units that can be quickly and easily identified, which minimizes confusion about site location. By including the parcel, land-use types, and land-use units on one map, most management plans can be completed on this map. The aerial image further supports identifying and locating specific features such as barns, shacks, boulders, pools, fences, brush, trails, and more.

Swanton Pacific Ranch's mission is to demonstrate and teach informed stewardship of the land, and this land-use map helps ranch managers accomplish this goal.

Improving watershed health and air quality in Washington, D.C.

Holli Howard, Mike Alonzo

Throughout the United States, more than 80 percent of the population lives in urban neighborhoods. Statistically, environmental issues and health risks are heightened in these areas. Studies have demonstrated that urban forests play a vital role in reversing many of the negative effects of city living. However, the urban environment is often tough on trees—pollution, weather extremes, neglect, traffic accidents, construction, and vandalism all create challenges for healthy trees to mature.

Washington, D.C., has a total area of 68.3 square miles, of which 10 percent is water. Trees are found in parks and yards, along roadways and paths, open green spaces, and undeveloped forests.

The goal of urban forest management in Washington is not to simply fill empty spaces, but to ensure that trees thrive to create a healthier environment for urban citizens. To reach this goal, planners and local residents work together to identify viable planting locations to maximize the benefits of urban trees, which include improving air quality and managing storm water runoff. To maximize the benefits of the urban forest, it is important to plant the right trees, in the right way, and in the right location.

Using the technology of GIS and spatial analysis, city planners can select appropriate planting sites for best results. Using maps that incorporate base layers such as roads and other impervious surfaces, along with watershed boundaries, stream layers, and IKONOS land-cover data, they can select optimal planting locations in riparian buffer zones, which maximizes the volume and quality of the city's tree canopy.

A visual solution

Roughly 40 percent of the District of Columbia's land cover is impervious surface made up of buildings, roads, sidewalks, and parking lots. The improved air quality and reduction of sewer system overflows resulting from planting a tree over an impervious surface is four times that of planting over a permeable surface such as grass. Therefore, urban land managers wanted to identify priority areas to plant where the tree canopy will cover an impervious surface.

The map's first area of focus is asphalt parking lots. Washington has more than 8,100 surface parking lots covering 2,100 acres or 5 percent of total land cover. During summer months, temperatures inside cars parked under shade trees are often 45°F cooler than those parked in full sun. Additionally, pavement that is in the shade can be up to 35°F cooler, which increases the life of asphalt and reduces maintenance costs. Overall, pavement cooled by shade trees reduces the urban heat island effect, making cities more livable and healthy.

The parking lots in D.C. generate more than two billion gallons of water runoff each year. This runoff contributes to flooding and sends pollutants like oil, sewage, and trash into the District's rivers. Tree canopy over parking lots slows and captures rainfall. Bioretention areas planted with trees reduce runoff, trap pollutants, and reduce sediments that flow into the District's waterways.

On warm days, gasoline vapors combined with air pollutants produce smog, triggering health alerts for an average of at least twenty days each year in the District. Far less gasoline evaporates from cars parked in the shade of trees. Lower temperatures from shaded parking lots means less smog for those most affected by high smog levels—children, the elderly, and asthmatics.

Maximizing the benefits of urban trees through strategic planting techniques

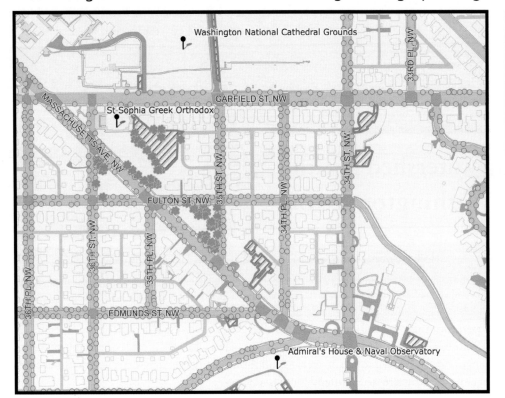

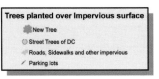

NAD Maryland State Plane 1983
GCS NA 1983

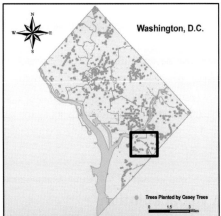

NAD Maryland State Plane 1983
GCS NA 1983

Data courtesy of Casey Trees/University of Vermont, USDA Forest Service; Casey Trees/Holli Howard; District of Columbia Geographic Information System.

The second area of focus on this map is riparian buffer zones surrounding stream and riverbeds. Two large rivers run through the District, the Anacostia and the Potomac. Forest cover is a critical component and indicator of a watershed's long-term ecological health and integrity. A 35-foot buffer on either side of a stream is generally considered the minimum width to provide shading, water quality, and other ecological benefits and is applied by both Maryland Department of Natural Resources and Prince George's County, Maryland, as the minimum riparian buffer required for preservation and/or reforestation in land development projects. Montgomery County employs a minimum 100-foot buffer with a variable width buffer system that factors in both stream sensitivity and steepness of valley side slopes. Currently, the District of Columbia does not have any stream buffer protection regulations.

Resource tables

Data dictionary

General data description	Data sources
D.C. infrastructure base data 2005	DCGIS-OCTO.
D.C. street tree locations 2002	Casey Trees.
D.C. land cover for Washington, D.C., 2006	IKONOS image purchased from GeoEYE.
IKONOS satellite image 1m resolution processed for land cover	Processed for land cover, specifically tree canopy, by the University of Vermont Spatial Analysis Lab.

Software dictionary

Software	Description
ESRI ArcGIS Desktop	ArcMap tools used for map display.
ArcGIS Spatial Analyst extension	Analysis tool to reclassify and calculate cell statistics.

Additional resources

Resource	Description and source
GIS manager or cartographer	Intermediate/advanced knowledge of ArcGIS, including raster analysis using the ArcGIS Spatial Analyst extension.

Recipe for map-building success

Casey Trees created this map by compiling and analyzing planting data and land-cover data for the Washington, D.C., area. The map highlights sites where trees can be planted to strategically combat urban environmental problems such as urban heat islands, poor air quality, and sewer system overflows.

Step 1: Compile existing data

The District of Columbia Geographic Information System (DCGIS) has an impressive collection of GIS data that is easily downloadable from its Web site. DCGIS GIS technicians digitized land features from aerial photos in 2005. Through analysis of this data, GIS specialists at Casey Trees generated GIS layers of impervious and permeable surfaces.

Casey Trees then used a street tree inventory data layer that it developed in 2002 to model a proposed tree canopy over the streetscape (roads, sidewalks, and intersections). This dataset spatially represents 130,000 street tree spaces in D.C., including the size and condition of the trees.

Step 2: Process IKONOS imagery

Casey Trees GIS specialists obtained high-resolution (1 meter) IKONOS satellite imagery taken of the D.C. area in July 2006 from GeoEye. Researchers at the Spatial Analysis Lab at the University of Vermont then classified the image to obtain the land-cover and tree canopy polygons.

Step 3: Identify priority planting areas

The District plants street trees an average distance of 40 feet apart. We used DCGIS impervious surface and boundary datasets (2005) to define the surface boundaries for the impervious or pervious base layer analysis, then performed both vector and raster analysis.

Based on average tree canopy, we created 20-foot buffers around parking lots to pinpoint the priority planting areas. Using the ArcGIS Spatial Analyst extension, we converted this vector dataset to raster in the same projection and 1-meter cell size, like the IKONOS imagery, to ensure a precise overlay. Each layer was reclassified and given a weight using the weighted overlay and raster calculator tools.

Step 4: Plant trees and collect data

For each tree planted, Casey Trees field teams identified the location and collected data such as species and condition. They used tablet PCs with a customized ArcMap form to record observations and measurements.

Step 5: Determine colors and symbols

In a vegetation analysis map or any land-cover map, colors should reflect the actual surface type, for example, using green for vegetation and blue for water. We wanted to show the areas in D.C. that were highlighted, as well as give some context to the capital city by showing the Washington National Cathedral grounds.

Conclusion

GIS is now an indispensable tool in long-range strategic land-use planning to analyze and set tree cover goals in Washington, D.C. To ensure that GIS helps set achievable tree cover goals, urban planners and managers need to understand the benefits of increased canopy cover, including reduced uncontrolled storm water runoff and better air quality. The methods we used here are based on models and data for the District of Columbia, but they are easily transferable to other municipalities interested in quantifying the relationship between tree canopy and environmental health issues.

By using GIS technologies, local organizations in Washington, D.C., have made a significant impact, not only in terms of collecting and sharing important data, but also in raising public awareness about the importance of trees in urban areas. Additionally, the data used for analyzing tree cover allows managers to quickly answer questions about the city's canopy, species mix, individual tree locations, and what trees provide in terms of health and other benefits.

In just five years, urban forest professionals in the nation's capital are leading by example, implementing state-of-the-art technology not only in its regreening efforts but also in developing a repeatable approach to achieving its regreening goal. Using GIS has facilitated new opportunities for strategic and operational tree program management within the city's infrastructure as well as its partner organizations.

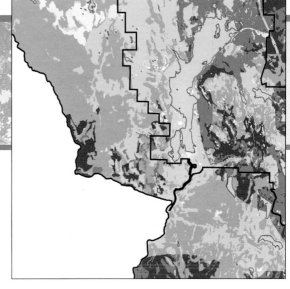

A wildfire risk management system for decision support

Bruce Blackwell, Franz Feigl

The past two decades have seen tremendous advances in the use of GIS and spatial modeling techniques to support fire management planning. A wildfire risk management system (WRMS) was recently adapted from the Australian Wildfire Threat Rating System for use in the province of British Columbia, Canada, and has since been applied by the authors in a number of different contexts and scales. One of the WRMS applications built was for the Capital Regional District (CRD), where the specific goal was to quantify the various facets of wildfire risk in a spatially explicit manner. We developed software to analyze the effects on water quality and other important watershed values, thereby providing the CRD with a wildfire risk management decision-support system.

Located on southern Vancouver Island, British Columbia, Canada, the CRD represents the regional government for thirteen municipalities and three electoral areas. The CRD's Water Services Department is the wholesale water supplier to core municipalities, including Victoria, the capital of British Columbia. The CRD maintains the Greater Victoria water supply area, about 11,000 ha in size and located west of Victoria in the Sooke Hills. The supply area contains a number of water reservoirs that fulfill the region's water needs and is off-limits to the public.

The CRD considers forest fires within the water supply area to be a significant risk to water quality, as fires may increase the amount of sediment, nutrients, and fecal coliform that enter the reservoirs. Sediments from surface erosion can create conditions that pose a threat to aquatic life, increased nutrient levels can trigger algae blooms that can affect the taste and odor of drinking water, and increased fecal coliforms levels can pose a direct threat to human health.

Historically, forest fires in the region range from less than 1 ha to 10,000 ha in size. Fire size and severity depends on current environment conditions and the ability of fire suppression resources to detect and suppress the fire. The CRD is particularly concerned that under conditions of extreme fire behavior or in the event of multiple fire starts, suppression resources be able to contain a given fire. Adding to this concern is the growing population of Greater Victoria and increasing development closer to the boundaries of water supply area, which heightens the risk associated with human-related fire ignitions, a major source of local forest fires.

A visual solution

The WRMS modeling software assesses a fire in terms of two factors, probability and consequence. The probability factor provides a measure of the likelihood of a fire, while the consequence factor provides a measure of the effects of a fire. The application makes use of the interaction between wildfire probability and associated consequences to determine the level of management intervention that may be required. For example, the most appropriate management strategy for an area rated low both in terms of probability and consequences might be "do nothing," while for an area rated low in terms of probability but high in terms of consequences it might be "develop emergency plans." On the other hand, if an area is rated high in terms of probability but low in terms of consequences, the most appropriate management strategy might be "monitor or mitigate." "Proactive intervention" is likely called for if both probability and consequence ratings are high.

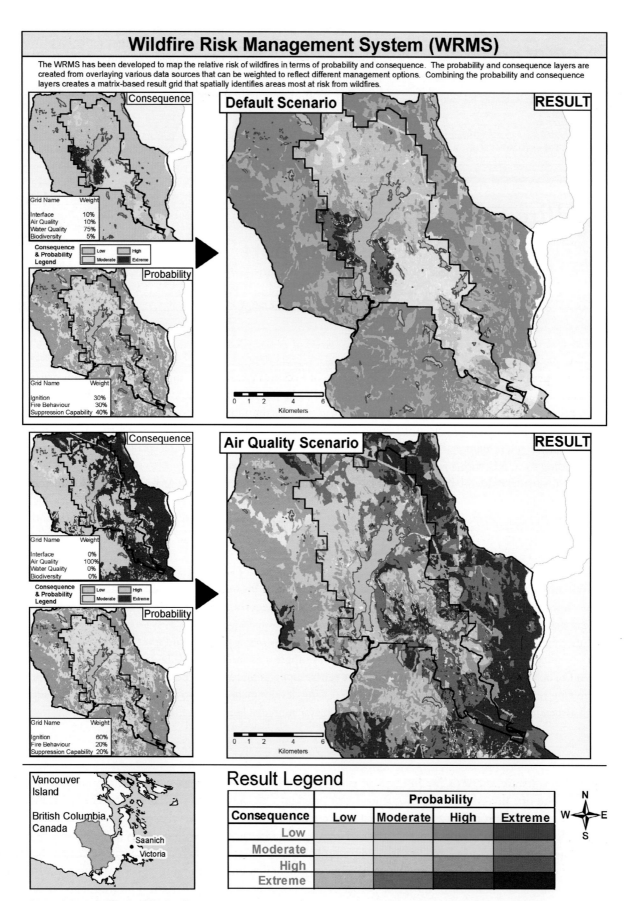

Both the probability and the consequence factor are computed from a number of components that may differ based on context, scale, and availability of local data sources. Components are composed of subcomponents, which represent distinct data layers.

Three components and the following subcomponents make up the probability factor:
- Probability of ignition: ignition potential, spatial fire distribution, and natural and human-caused fires
- Fire behavior: fire intensity, rate of spread, and crown fraction burned
- Suppression capability: constraints to detection, proximity to roads, proximity to water sources, water delivery constraints, helicopter arrival time, air tanker arrival time, and terrain steepness

Four components, which reflect the specific concerns of the CRD, and their subcomponents make up the consequence factor:
- Urban interface: key infrastructure, nonresidential values, and residential values
- Air quality: air-shed sensitivity rating, smoke venting potential, smoke production potential, proximity to population centers
- Water quality: turbidity, nutrient load, color, taste and odor, proximity to reservoir, proportion of water supply area, importance to water supply area
- Biodiversity: CDC element occurrences, red and blue listed elements

As an add-on for CRD Water Services, their WRMS application is tied into the Canadian fire weather behavior system, a system that generates daily fire weather indices (temperature, relative humidity, precipitation, and wind speed). Hence, CRD Water Services is able to produce fully spatial wildfire risk predictions daily (it takes about five minutes to create a scenario), and the department uses this resource to provide staff with very specific and spatially explicit instructions.

CRD GIS specialists are able to modify assumptions and create various scenarios by changing the relative weights of both probability and consequence subcomponents and/or associated components. Relative weights represent the subcomponents' and components' relative importance in the planning process, and are entered as values between 0 and 1. For example, the default scenario assigns a weight of 0.75 to the water quality component, which reflects the mandate of CRD water services. Upon setting the relative weights, the application hierarchically compiles first respective subcomponents into their associated component layers, and then all components into the associated factor layers. The process culminates with the creation of both a probability and a consequence layer, which in turn are combined into a single results layer that reflects the four management strategies outlined.

Resource tables

Data dictionary

General data description	Data sources
British Columbia (BC) boundary	Integrated Land Management Bureau (ILMB).
DEM	Integrated Land Management Bureau (ILMB), BC Terrain Resource Information Management (TRIM) data.
Forest resource inventory, roads, and utilities	Capital Regional District.
Consequence layer, probability layer, and results layer	Output of WRMS modeling software.

Software dictionary

Software	Description
ESRI ArcGIS Desktop	Multiple data frames used, color rendering, transparencies, labeling, layer files, and geoprocessing.
Microsoft PowerPoint	Used to make custom legend.
Wildfire Risk Management System (WRMS) [proprietary model]	Modeling software used to generate values for probability, consequence, and result layers.

Recipe for map-building success

The map presented here depicts the results of a WRMS analysis for CRD water services for two scenarios, namely the default scenario and the air quality scenario. The map is not typical of the cartographic output as per CRD water services guidelines. Rather, the map has been specifically compiled for this publication to highlight the results of a wildfire risk assessment and the impact of changing management assumptions by selecting different relative weights.

Step 1: Establish the need and format of the WRMS

This is a critical step, and involves establishing the exact reasons why a client wants to have a WRMS built. Our experience has taught us that this is usually directly tied to the client's mission statement. In the case of the CRD water services, the purpose for the WRMS was to help develop a wildfire risk management plan for the Greater Victoria water supply area. This step also involves identifying desired component and subcomponents, as well as creating individual subcomponent lookup tables that will be used to break up the subcomponent layers into various categories.

Step 2: Collect raw data

The kind of raw data collected is a function of the purpose for which the WRMS is used. In Canada, large amounts of data are held in the public domain and are therefore relatively easy to obtain. Data of a more specialized or sensitive nature might only be available from public or private corporations, municipalities, universities, and nongovernment organizations. On occasion, desired data might not be readily available in a format suitable for cost-effective inclusion in the model.

Examples of GIS-ready public domain data in British Columbia are data collected under the Terrain Resource Information Management (TRIM) program, such as a digital elevation model (DEM) or hydrological inventories. Forest inventory data, such as the Vegetation Resource Inventory (VRI) is generally in the public domain as well; however, some corporations and municipal governments have developed their own inventories. Examples of nonpublic domain data are helicopter and air tanker base locations, water intake locations, and infrastructure types and locations. In addition, not all data is spatial: examples are helicopter and air tanker traveling speeds with and without loads, response times of fire fighting crews to reports of a fire, or lists of human causes of forest fires (campfires, arson, carelessness, etc.).

Step 3: Generate derivative data

This step involves generating derivative data, which is the creation of new data from existing data. No standardized approach exists because the generation of the derivative data depends on the nature and the quality of the existing data, as well as on the client's specifications. However, here we provide two examples from the CRD: we created the helicopter and air tanker arrival time subcomponents by generating circular buffers around helicopter and air tanker locations, based on take-off time, travel speeds, and travel range. The creation of the probability surfaces for human- and lightning-caused fires subcomponents requires more sophisticated data generation and data processing routines. Detailed incident lists, complete with spatial references were provided by the client in spreadsheet format, and we wrote customized applications to generate the appropriate subcomponents from this information.

Step 4: Stratify data layer (subcomponents)

After all subcomponents have been generated, they are stratified based on the subcomponent lookup tables generated in step 1. For example, the air tanker and helicopter arrival times might be stratified by three-minute intervals, or the lightning-caused fire probability surface by the number of lightning strikes by square kilometer per decade. Generally, relative weights ranging from 0 to 10 are assigned to the various classes (the WRMS allows the user to change these values). After that, the subcomponents are converted to raster and loaded into the WRMS.

Conclusion

The WRMS is a cost-effective application that provides the user with a quick way to calculate and map spatially explicit wildfire risk. Using average conditions, the output can be used to develop a wildfire risk management plan. Because the model can be tied into the Canadian fire weather behavior system (or any similar system that tracks daily fire weather indices), the wildfire risk can be calculated on a daily basis. This provides management with a tool to spatially fine-tune decision making (for example, allowing open campfires in certain locations but not in others). Each application is custom-built and reflects specific needs; but since the model is also designed for gaming, the user therefore is able to consider different scenarios.

A potential drawback of the system is that it relies on input data that might be somewhat outdated. Unfortunately, some data is almost certainly outdated even while it is being compiled, such as forest inventories and data reflecting growth, such as population growth. The use of outdated data is a common problem in GIS, and, in our opinion, a problem currently still impossible to overcome. However, as logging is not permitted in the Greater Victoria water supply area, fire behavior (i.e., fire intensity, rate of spread, and crown fraction burned) in the CRD is subject only to disturbances such as wind throw and fire. Of course, the CRD is addressing the latter with the WRMS.

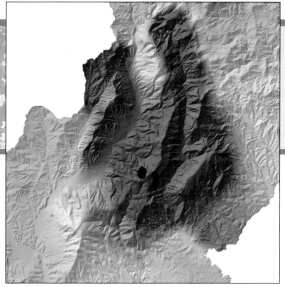

Mapping southern pine beetle hazard in the Pisgah National Forest, North Carolina

Weimin Xi, Lei Wang, Maria D. Tchakerian, Robert N. Coulson

Established in 1916, the Pisgah National Forest is one of the first national forests in the eastern United States. Covering more than 500,000 acres, the Pisgah features rugged mountains, waterfalls, rivers, and wilderness areas. The Pisgah is known as the "Cradle of Forestry" since it is the site of the first school of forestry in the United States, the Biltmore Forest School. This school operated during the late 1800s and early 1900s and was originally managed by Gifford Pinchot, the first chief of the U.S. Forest Service.

Between 2000 and 2002, the southern pine beetle created catastrophic damage to a wide variety of pine species found throughout forests in the southern Appalachian Mountains. The southern pine beetle (*Dendroctonus frontalis*) is the most destructive insect pest of pine forests in the southeastern United States. Tree species affected include Table Mountain pine (*Pinus pungens*), pitch pine (*Pinus rigida*), shortleaf pine (*Pinus echinata*), and Virginia pine (*Pinus virginiana*).

This map was created to identify areas susceptible to future southern pine beetle (SPB) outbreaks in the Grandfather Ranger District of the Pisgah National Forest in western North Carolina. Created to provide forest owners and managers with spatial information on SPB hazard, the map shows specific high-hazard areas where prevention and restoration practices should be conducted. Additionally, this map provides spatial information on SPB hazard for Texas A&M University's ongoing spatial modeling project. The project investigates the utility of landscape models as a decision-making tool for predamage impact analysis and post-damage restoration planning.

A visual solution

Using information that is accessible to the public, this map indicates areas that are most susceptible to potential attack showing a landscape-level SPB hazard rating. Forest owners and managers can use this information to increase the effectiveness of forest management planning and assessment in order to suppress and prevent future SPB outbreaks.

To create the map, research scientists at Texas A&M University developed a procedure that integrates spatial statistics techniques along with scientific knowledge of key factors conducive to SPB activity. This approach is widely applicable in the southern Appalachian Mountains and more generally to adjacent areas of the southeastern United States.

The hazard rating used in this map is based on the spatial configuration of the yellow pine hosts located throughout the landscape with high-hazard areas identified on the map in red. Area-based spatial clustering analysis was used to generate the Getis-Ord general G index as the indicator of susceptible hotspots to SPB. Getis-Ord is a clustering tool used to measure the concentration of high or low values for a specific study area. The tool calculates both the predicted and expected results of the analysis, which is helpful for identifying unexpected spikes in the data.

In addition, elevation and the locations of SPB outbreaks in year 2000 are presented to help forest owners and managers prioritize areas where they should conduct prevention activities.

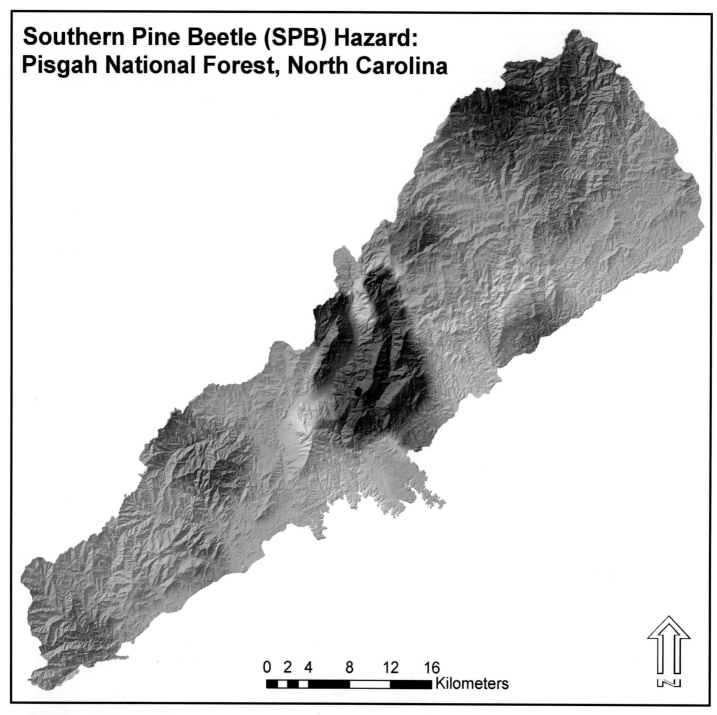

Data courtesy of U.S. Geological Survey; the Southern Appalachian Man and the Biosphere (SAMAB) Program; the National Forests in North Carolina; Weimin Xi; Lei Wang; Maria D. Tchakerian; Robert N. Coulson.

Resource tables

Data dictionary

General data description	Data sources
Digital elevation models (DEM)	U.S. Geological Survey.
Forest stands	The Southern Appalachian Assessment Database.
Southern pine beetle infestations	National Forests in North Carolina.

Software dictionary

Software	Description
ESRI ArcGIS Desktop	Georeference tool, Editor tool, ArcCatalog data import tool, and projection tool.
ArcGIS 3D Analyst extension	Terrain Modeling.
ArcGIS Spatial Analyst extension	Neighborhood Analysis tool.

Additional resources

Resource	Description and source
Satellite image	U.S. Geological Survey.

Recipe for map-building success

Step 1: Obtain data

Download forest stand and digital elevation model (DEM) data from publicly accessible sources. SPB infestation locations and boundaries for National Forests in North Carolina are available online in PDF format. These files are then digitized and spatially georeferenced.

Step 2: Forest stand classification

Reclassify the forest stand types into percentage of pine trees according to the description of each stand type. For example, forest stands that belong to pine dominated types will have pine tree coverage at a possible percentage from 70 to 100 percent.

Step 3: Spatial clustering analysis

Run the area-based spatial clustering analysis tool. We chose the Getis-Ord G index calculator in the ArcGIS Spatial Statistics toolbox. The Spatial Statistics toolbox, available with ArcGIS Desktop, contains statistical tools for analyzing the distribution of geographic features. The Getis-Ord p-value (probability value) from this analysis quantifies the spatial clustering tendency of pine trees. It is used as the indicator of SPB hazard rate.

In the ArcGIS Spatial Analyst extension, the neighborhood analysis tools are used to smooth the output data. We used a kernel size of 5x5 pixels.

Step 4: Create shaded relief map

Derive the shaded relief map from the DEM by using the hillshade function in ArcGIS 3D Analyst.

Step 5: Load SPB data

Georeference the SPB infestation data obtained from the national forests in North Carolina to the GIS environment. Additional SPB outbreaks in the year 2000 locations were obtained from hard-copy maps and then digitized.

Step 6: Organize layers

The Getis-Ord G index layer received priority over the hillshade layer in ArcMap. The SPB outbreak data was placed as the topmost layer so it could be quickly and easily identified.

Step 7: Create map

A color scheme changing from blue to red was chosen for the G index. A 50 percent transparency was applied to show the underlying terrain features depicted by the shaded relief map. Other cartographic elements such as a legend, index map, and scale bar complete the map.

Conclusion

This map was developed using spatial analysis techniques and scientific knowledge of southern pine beetle (SPB) population dynamics, including relationships between hosts and SPB spread patterns. The spatial distribution patterns of SPB hosts in the landscape were synthesized to indicate the degree of SPB hazard using outbreak data from the year 2000 to validate analysis results. This approach helped to reveal the landscape-scale SPB hazard in the Grandfather Ranger District of the Pisgah National Forest.

The map is clear and precise, which enables nontechnical personnel or the general public to easily interpret findings. Research scientists at Texas A&M University chose a color legend that gives the strongest contrast between areas with different SPB hazard ratings. The red color represents hot spots, the main theme of the map. An insert map showing the location of our study area is also necessary to orient the reader; here the boundary of western North Carolina is used for the small insertion.

A hillshade relief map was chosen as the basemap layer because it provides the best combination of spatial reference information and visualization effects. The locations of previous SPB outbreaks are identified to demonstrate the validation of the model: these hot spots are located where the hazard rate is expected to be highest. A graduated color scheme from blue to red representing the hazard rating provides the best visualization of the spatial patterns of SPB hazard.

This map uses data sources that are all in the public domain. It does not require complex computations and all tools used are available in ArcGIS Desktop.

Predicting land-cover change and forest risk in the Bolivian lowlands

Clark Labs

Scientific evidence suggests that land change is expected to be the most significant contributor to biodiversity loss in the twenty-first century. In part, this is a result of changes related to habitat degradation, fragmentation, and destruction. Due to tremendous economic growth in recent decades, the Bolivian lowlands experienced forest loss of almost 3 million hectares between 1992 and 2004, with almost half of this loss occurring between 2001 and 2004.

Conservation International, an organization at the forefront of natural systems preservation, uses GIS technology to study historical habitat loss and to predict future change. Clark Labs collaborated with Conservation International to develop a GIS software solution called the Land Change Modeler that projects future land-cover change and assesses its effects.

This map predicts vulnerable areas in the Bolivian lowlands resulting from deforestation and land-cover change to the year 2015. The map enables Conservation International to make better and more informed resource planning decisions and prioritize locations for further evaluation to reduce the potential impacts of land-cover change. This map is important to planners and the conservation community so they can better manage an evolving landscape.

A visual solution

Bolivia is a landlocked country, bounded by Brazil to the north and east, Paraguay to the southeast, Argentina to the south, Chile to the southwest, and Peru to the northwest. With an area of 1,098,580 square kilometers, Bolivia is composed of three geographic zones: the mountains and highland plateau (Altiplano) in the west, the semitropical Yungas and temperate valleys of the eastern mountain slopes, and the tropical lowlands of the Amazon basin. The lowlands comprise approximately 70 percent of Bolivia's total area.

Forests dominate more than 65 percent of Bolivia's lowland land cover, and scientists working on this project consider understanding the land's changing dynamic to be a critical component to developing policies and legislation that ensure Bolivia's forests are sustainably managed. Since researchers now understand that tropical forests are an essential component for maintaining the functioning of the global climate system and that forests also provide habitat for more than half of the world's species, protecting what is left of Bolivia's natural forest ecosystems is crucial to the world's survival.

The map identifies areas most vulnerable to degradation, assuming continuing trends of current human behavior. Prediction maps typically convey a single scenario in time, but soft predictions such as those the model used to create this map demonstrate a land cover's *potential* to change at a future date. This is a useful and realistic approach for identifying areas of high risk and facilitates the identification of biodiversity hot spots for prioritizing conservation areas.

Clark Labs analyzed land-cover maps for the Bolivian lowlands for 1992, 2001, and 2004 to create the prediction. Geospatial analysts applied a set of twenty-eight static and dynamic driver variables that included proximity to infrastructure, slope, soil type, and others to make predictions. Information on known road improvements for Bolivia was also used to further refine the model.

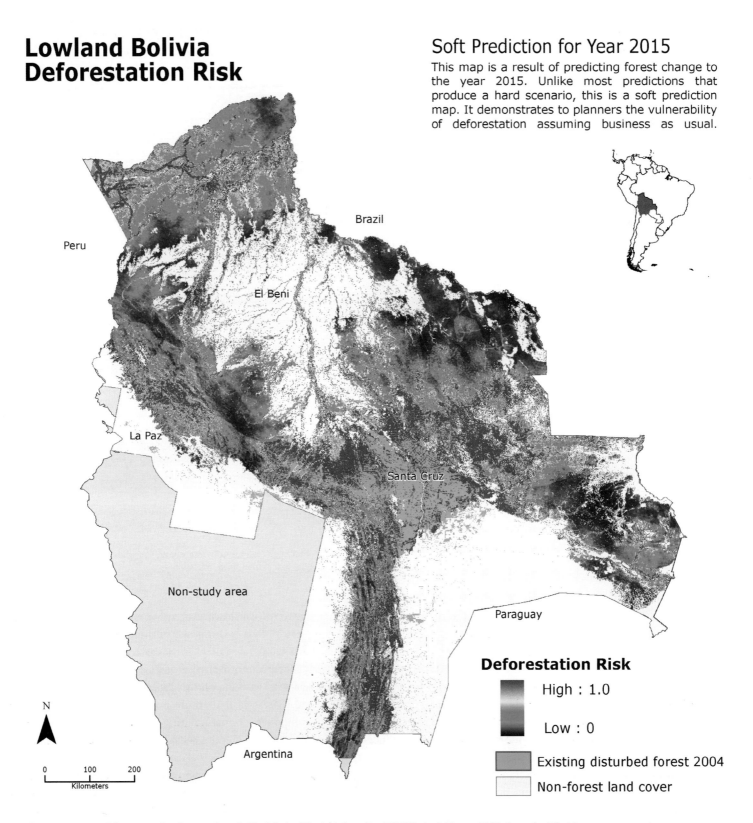

The Geography Department at the Noel Kempff Mercado Natural History Museum in Santa Cruz, Bolivia, developed the maps. The museum received support from Conservation International, World Wildlife Fund, The Nature Conservancy, and the Bolivian government to perform activities related to the analysis. Research staff at Clark Labs developed the model.

Resource tables

Data dictionary

General data description	Data sources
Land-cover maps over time	Typically classified from satellite imagery, but can be acquired from some state or local government agencies.
Environmental driver variables	State and local agencies, remotely sensed data, GIS derivatives.

Software dictionary

Software	Description
ESRI ArcGIS Desktop	Development of GIS derivatives and environmental driver variables.
ArcGIS Spatial Analyst extension	Used for raster-based spatial modeling and analysis.
Land Change Modeler for ArcGIS	Change analysis, transition potential modeling, and change prediction.

Recipe for map-building success

Step 1: Obtain data

The first step to developing this predictive map is to obtain three classified raster land-cover maps, all spanning different time periods.

Step 2: Identify transitions of interest

Specify a specific transition type that is of interest and requires further study, such as a transition between natural forest ecosystems to developed properties.

Identify the variables that drive the type of transitions taking place. Environmental variable maps must be supplied to uncover the inherent suitability of land that will undergo transition. For instance, if we want to determine potential environmental changes caused by new development, we may consider the slope of the terrain, distance to water sources, distance to roads, and distance to previously developed land.

Step 3: Uncover past land-change dynamic

Clark Labs performed a change analysis to better understand where changes are taking place. Land-cover maps from 1992 (time 1) and 2001 (time 2) were compared using the Land Change Modeler tool.

Step 4: Create and calibrate the model

To explain the land-cover change in Bolivia from 1992 to 2001, we used environmental driver maps as input to empirically derive the model and predict a future scenario map. This scenario map was then compared to the 2004 (time 3) land-cover map for Bolivia to validate the accuracy of the model and verify whether the correct driver variables were selected to explain the change. The accuracy is determined to be low or high by comparing the time 3 map produced by the model to the map of actual change in 2004.

Calibration of the model to an actual map is critical to evaluate model accuracy. If the model produces a map that compares poorly to the map of the third date, other environmental driver variables may need to be identified and existing ones reexamined.

Once the model is sufficiently calibrated, we can produce a number of future prediction scenarios. The model will predict, for the specified future date, the allocation of land-cover change.

Step 5: Create map

We used ArcMap to create a high-quality cartographic output with model results that can be easily interpreted and communicated to a nontechnical audience.

Conclusion

This map of forest vulnerability is important to demonstrate that one can empirically derive future land-cover scenarios from known environmental factors, allowing for a rapid description of land-cover change and increasing our understanding of environmental effects on the landscape.

Additionally, this map visually presents a comprehensive assessment of change potential in Bolivia, which is understood by GIScience modelers as a "soft" prediction output. It illustrates vulnerability to deforestation rather than the more common single realization or "hard" prediction. Instead of committing to a specific change scenario, the soft prediction output does not forecast what will change, but rather presents the degree to which the areas have the right conditions to precipitate change. Soft prediction acknowledges the reality of a dynamic environment, increasing its utility and providing an indication to planners and conservation practitioners of potential hot spots of future change.

For this map, it was important that the areas of high risk for deforestation be easily discernable. To achieve this, we decided to visualize only the forest land-cover change potential, thereby making it easier for planners to assess the effects. Since the data is meant to show two opposing situations, a diverging color scheme was used to enhance the contrast between areas of high and low risk to deforestation.

Prioritizing restoration in fire-adapted forest systems

Chris Zanger, Amy Waltz

CHAPTER 17

Forests and woodlands in Oregon and the northwest United States are burning in ever-increasingly severe and intense wildfires. Despite work by federal agencies to reduce fuels and restore these systems through prescribed fire, thinning, and wildland fire use, conditions are getting worse, not better. The size of fires ignited both naturally by lightning and by humans is increasing in acreage and burning at severities and intensities much higher than the range of natural variability for many of our forested systems.

Stakeholders, including wildlife and conservation interests, have expressed concerns and at times opposed agency management of forests. Specific concerns that often make reaching consensus difficult include which location to treat and how much of the landscape to treat, prescriptions for restoration, treatment methods, and cumulative treatment effects on wildlife habitat. Still, despite differences, most of these stakeholders support a cooperative effort to find proactive solutions.

The Upper Deschutes Basin Fire Learning Network (FLN) created this map as a base layer to identify on-the-ground projects that advance the restoration of fire-adapted systems. The Restoration Priority map highlights forest stands historically adapted to frequent fire that today support very different forest structure due to fire exclusion and other management choices. Our 2 million-acre landscape includes U.S. National Forest, Bureau of Land Management, state, and private lands. This map has been successful showing the scale at which restoration opportunities are available.

A visual solution

Long-standing disagreements over how our forests should be managed to conserve sustainable ecosystems while meeting multiple resource goals can result in little action taken in places most threatened by uncharacteristic fire. Other factors currently limiting restoration action include declining federal budgets to plan and implement forest restoration and a lack of infrastructure (mills, cogeneration, and so on) to remove and process small diameter timber and timber by-products of restoration treatments.

The FLN is a collaborative formed to work toward consensus for forest management on federal landscapes using the best available science. Multiple stakeholders, including federal agencies, state agencies, environmental and conservation groups, timber companies, and community groups work together to develop areas of consensus and open analyses of the multiple values in this area.

Prioritized restoration stands selected for this map were highlighted through an overlay and analysis of
- ponderosa pine and mixed conifer stands (fire regimes I and III, map 2);
- stand structural data resembling second growth stands (mid-successional stage, map 3); and
- stands with closed canopy (greater than 40 percent, map 4).

Mid-successional, closed-canopy stands exist in much higher proportions today than were found historically. Thinning small-diameter trees and returning fire to these stands would move the stands from a closed, mid-successional stage to an open, mid-successional stage, which would lower the departure from the natural range of variation for these vegetation types and set these stands on a trajectory to create more open late-successional stands. This map does not suggest that all these acres should be treated; however, treating a subset of these stands would move the forest toward the stand structure found prior to alteration of the fire regime.

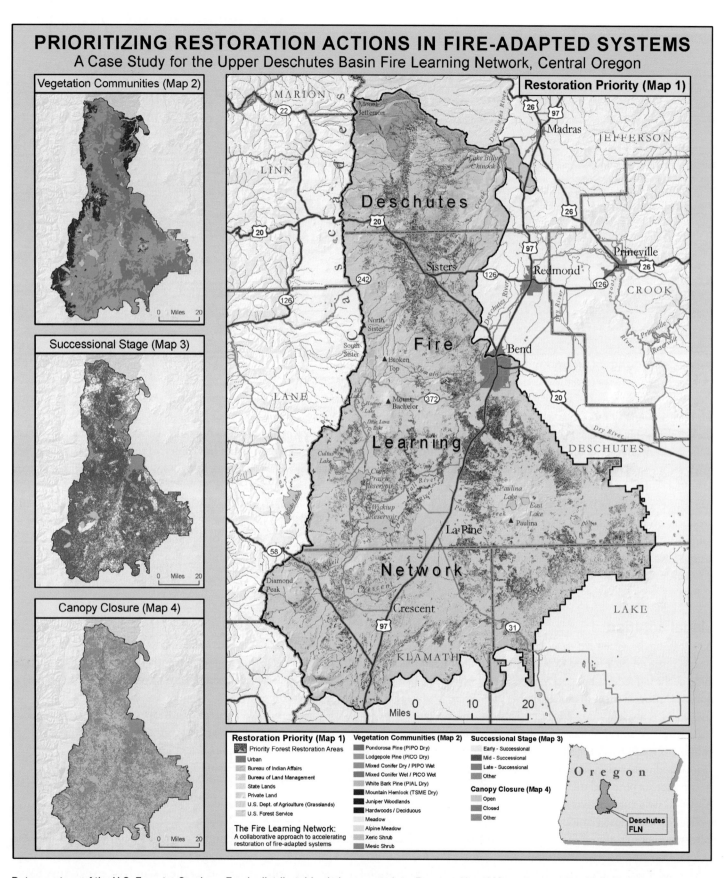

Resource tables

Data dictionary

General data description	Data sources
Landownership	Bureau of Land Management, Prineville District, Oregon.
Biophysical setting	Raster map of potential vegetation with disturbance for the landscape. The Upper Deschutes Fire Learning Network developed this with a local interagency technical team from a Forest Service potential vegetation layer. Some national forests have this data available or LANDFIRE data is available for free for the whole country from the LANDFIRE Web site.
Current vegetation communities	Raster map of the landscape. A 2004 LANDSAT imagery dataset with forest structural classifications was available from the Deschutes National Forest. Some national forests have this data available or landfire data is available for free for the whole country from the LANDFIRE Web site.
Canopy closure	Raster map of overstory canopy of closure. A 2004 LANDSAT imagery dataset with forest structural classifications was available from the Deschutes National Forest. Some national forests have this data available or landfire data is available for free for the whole country from the LANDFIRE Web slte.
Successional stage	Raster map of dominate overstory age class landscape. We contracted a cross-walk of the forest structural data to a five-box succession model (state and transition model). Some national forests have this data available or landfire data is available for free for the whole country from the LANDFIRE Web site.
Basemap data layers (road, cities, etc.)	Oregon Geospatial Data Clearinghouse, State of Oregon.

Software dictionary

Software	Description
ESRI ArcGIS Desktop	ModelBuilder and general cartographic analysis.
Fire Regime Condition Class (FRCC) mapping tool	This tool analyzes specific input grids and creates a summary result grid of FRCC and successional stage relative amounts as well as a Microsoft Access database of summary statistics to explain departure of current forest conditions from historic forest conditions. Available on the Fire.org Web site.

Additional resources

Resource	Description and source
Citation	Hann, W., D. Havlina, A. Shlisky, et al. 2003. Interagency and The Nature Conservancy fire regime condition class Web site. USDA Forest Service, U.S. Department of the Interior, The Nature Conservancy, and Systems for Environmental Management.

Recipe for map-building success

The restoration priority map (map 1) became an important layer in FLN's collaborative process for restoring fire-adapted systems and was the baseline for our analysis. Our next steps as a collaborative are to accumulate the spatial data or spatial surrogates for the multiple values our stakeholders treasure in this landscape. We are engaged in an iterative process to accumulate, overlay, and weigh multiple values to identify zones of agreement or consensus on the landscape.

The steps to create the restoration priority map (map 1) are as follows:

Step 1: Acquire base data

Acquire base data to examine ecological departure of current stand conditions from historical stand conditions. We used a standardized methodology developed for national assessment of ecological systems—the Fire Regime Condition Class[1] (FRCC). An automated ArcMap extension tool, the FRCC mapping tool is available from LANDFIRE to run a departure index based on data layers, including biophysical setting, current vegetation, and canopy closure. In creating a similar priority map, if these layers are not available locally for projects in the United States, course-scale data is available nationally from the LANDFIRE Web site. Our data layers were developed with a collaborative technical team (Forest Service, Bureau of Land Management, U.S. Fish and Wildlife Service, Oregon Department of Fish and Wildlife, and The Nature Conservancy).

Data used for analysis included the following:
- Biophysical setting (potential vegetation with disturbance): modified from a plant association group map, Deschutes National Forest.
- Current vegetation: we used a 2004 Landsat imagery dataset, Deschutes National Forest.
- Successional stage: this dataset was developed from a cross-walk of forest structural data to a five-box succession (state and transition) model. (Contracted with Spatial Solutions Inc.)
- Canopy closure: we used a 2004 Landsat imagery dataset, Deschutes National Forest.
- Nonspatial data: historical successional stage distributions. Biophysical systems were modeled with natural disturbances to determine the proportion of successional stages on the landscape within the natural range of variability (Vegetation Dynamics Development Tool).

Step 2: Analyze successional stage distributions

Use FRCC Mapping Tool (available online) to analyze how current successional stage distributions depart from the historical, reported as the Stand FRCC and Successional Stage Relative Abundance layers.

Step 3: Create multiple grid query

Use ModelBuilder and ArcGIS Spatial Analyst extension to create a multiple grid query to select the values of interest from the input grids and FRCC analysis results grids and create a new binary grid illustrating areas where all the pixels fulfill the grid argument. We selected stand characteristics that are currently drastically different from historical forest characteristics based on FRCC analyses and are also ponderosa pine or mixed conifer dry within a mid-successional stage and with a closed canopy.

Step 4: Create map

Develop a quality map based on cartographic principles.

Conclusion

In 2005, Fire Learning Network workshops examined the barriers to setting landscape-level management priorities based on ecological indices. These barriers included a lack of consistent data across ownership boundaries and across administrative boundaries within federal agencies. This map offered a new look at the landscape need for forest restoration. Creating this map allowed our collaborative group to successfully engage with the federal land management agencies, broaden our stakeholder group and build credibility with our partners.

The unique contribution of this map was to combine spatial layers commonly used by land managers to characterize problem areas. This particular display (maps 1 through 4) allows users and partners to assimilate multiple layers so they can understand a complex message across a very large landscape. Our collaborative continues to accumulate spatial data layers representing additional values, such as community protection, fire risk, wildlife habitat, recreation, cultural resources, and forest-product use. Here we feature this display because of its success and appeal to natural resource managers who assess and plan at landscape levels as well as its appeal to nonspecialist community stakeholders.

[1] Hann, W., D. Havlina, A. Shlisky, et al. 2003. Interagency and The Nature Conservancy fire regime condition class Web site. USDA Forest Service, U.S. Department of the Interior, The Nature Conservancy, and Systems for Environmental Management.

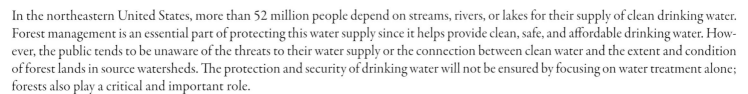

Protecting the drinking water supply in the northeastern United States

Rebecca Lilja

In the northeastern United States, more than 52 million people depend on streams, rivers, or lakes for their supply of clean drinking water. Forest management is an essential part of protecting this water supply since it helps provide clean, safe, and affordable drinking water. However, the public tends to be unaware of the threats to their water supply or the connection between clean water and the extent and condition of forest lands in source watersheds. The protection and security of drinking water will not be ensured by focusing on water treatment alone; forests also play a critical and important role.

The purpose of this map is to illustrate the direct geographic connection between forests, water, and people—a connection sometimes called the "forest-to-faucet" relationship. This map demonstrates the importance of private forests for protecting surface drinking water quality and illustrates the potential threats to those forests that will affect water quality. By analyzing these relationships on a landscape level, we can better determine priorities for management action and mitigation.

A visual solution

GIS analysts created a watershed condition index map, then compared it with watersheds across the northeastern United States that are able to produce clean drinking water. Water and land managers use the results of this analysis to guide and support conservation approaches and prioritize the most threatened areas. This map also serves to heighten awareness for local and regional policy makers and the public, helping them understand our dependence on forests for clean water. Through education, this map helps to motivate citizens, water utilities, local, state and federal government agencies and commissions, nongovernmental organizations, private sector firms, and educational institutions working together on a shared vision to increase conservation of forest lands in priority watershed.

Through a four-step GIS assessment, the U.S. Department of Agriculture (USDA), Forest Service created a watershed condition index map for the northeastern area based on nine physical and biological attributes. Using an overlay process, GIS analysts merged six data layers, including forest land, agricultural land, riparian forest cover, soil erodibility, road density, and housing density. The resultant layer produced an index that was used to compare the ability of watersheds across the twenty-state northeastern area to produce clean drinking water. Steps two through four of the analysis quantifies the magnitude and scope of forest-dependent drinking water supplies and identifies watersheds that are threatened by land-use change or that need to be managed to sustain and improve forest conditions for water supply protection.

Important Watersheds* at Risk of Future Forest Conversion

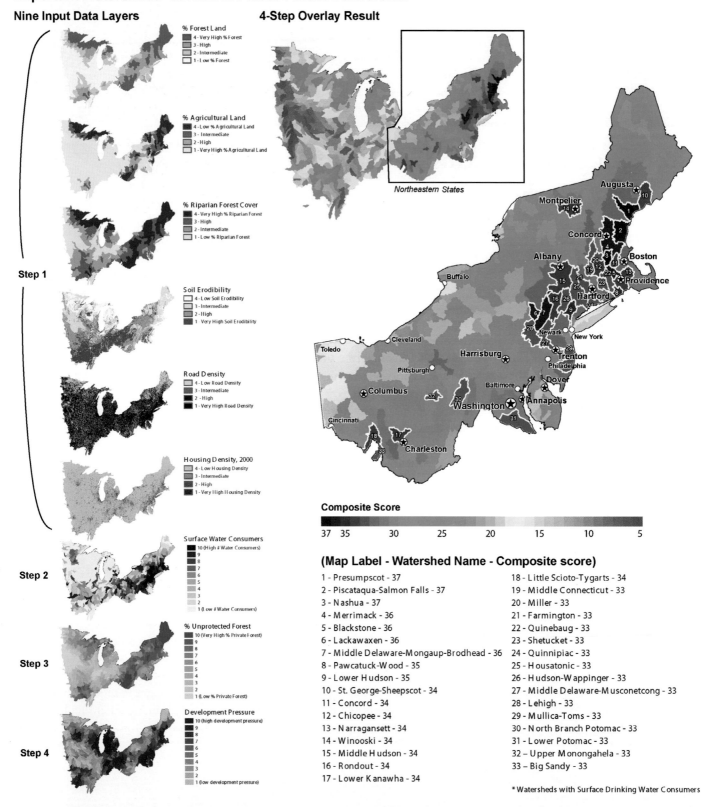

Nine Input Data Layers — Step 1: % Forest Land, % Agricultural Land, % Riparian Forest Cover, Soil Erodibility; Step 2: Road Density, Housing Density 2000, Surface Water Consumers; Step 3: % Unprotected Forest; Step 4: Development Pressure

4-Step Overlay Result

Composite Score: 37 35 30 25 20 15 10 5

(Map Label - Watershed Name - Composite score)

1 - Presumpscot - 37
2 - Piscataqua-Salmon Falls - 37
3 - Nashua - 37
4 - Merrimack - 36
5 - Blackstone - 36
6 - Lackawaxen - 36
7 - Middle Delaware-Mongaup-Brodhead - 36
8 - Pawcatuck-Wood - 35
9 - Lower Hudson - 35
10 - St. George-Sheepscot - 34
11 - Concord - 34
12 - Chicopee - 34
13 - Narragansett - 34
14 - Winooski - 34
15 - Middle Hudson - 34
16 - Rondout - 34
17 - Lower Kanawha - 34
18 - Little Scioto-Tygarts - 34
19 - Middle Connecticut - 33
20 - Miller - 33
21 - Farmington - 33
22 - Quinebaug - 33
23 - Shetucket - 33
24 - Quinnipiac - 33
25 - Housatonic - 33
26 - Hudson-Wappinger - 33
27 - Middle Delaware-Musconetcong - 33
28 - Lehigh - 33
29 - Mullica-Toms - 33
30 - North Branch Potomac - 33
31 - Lower Potomac - 33
32 - Upper Monongahela - 33
33 - Big Sandy - 33

*Watersheds with Surface Drinking Water Consumers

Data courtesy of U.S. Department of Transportation 2002. Bureau of Transportation Statistics (BTS): Roads; Natural Resources Conservation Service, Miller, D. A., R. S. White 1998. STATSGO: A conterminous United States multilayer soil characteristics dataset for regional climate and hydrology modeling; Lilja, Rebecca. 2006. Private Forests. Durham, N.H.: U.S. Department of Agriculture, Forest Service, Northeastern Area [unpublished dataset]; U.S. Geological Survey. 1999. 1992 National Land Cover Dataset; Hatfield, Mark. 2005. 30m Buffer of the 1992 National Hydrography Dataset (NHD). St. Paul, Minn.: U.S. Department of Agriculture, Forest Service, Northern Research Station [unpublished dataset].

Resource tables

Data dictionary

General data description	Data sources
Forest land	U.S. Geological Survey. 1999. 1992 National Land Cover Dataset.
Agricultural land by watershed	U.S. Geological Survey. 1999. 1992 National Land Cover Dataset.
Riparian forest cover by watershed	Hatfield, Mark. 2005. 30m buffer of the 1999 National Hydrography Dataset (NHD). St. Paul, Minn.: USDA Forest Service, Northern Research Station. [unpublished dataset] and U.S. Geological Survey. 1999. 1992 National Land Cover Dataset.
Road density	U.S. Department of Transportation. 2002. Bureau of Transportation Statistics (BTS).
Soil erodibility, k factor	Natural Resources Conservation Service, D. A. Miller, and R. A. White. 1998. STATSGO: A conterminous United States multilayer soil characteristics data set for regional climate and hydrology modeling.
Housing density by watershed	Theobald, David M. 2004. Housing density in 2000 [Digital Data]. Natural Resource Ecology Lab, Colorado State University. Fort Collins, Colorado.
Surface water consumers	U.S. Environmental Protection Agency. 2005. Safe Drinking Water Information System (SDWIS) Public drinking water system (PWS) consumers by 8-digit HUC.
Private forest by watershed	Protected Areas Database, Version 4. Conservation Biology Institute 2006; Wisconsin Stewardship Data; U.S. Geological Survey, Upper Midwest Environmental Sciences Center, 2005.
Development pressure per unit area	Theobald, David M. 2004. Housing density in 2000 and 2030 [Digital Data]. Natural Resource Ecology Lab, Colorado State University. Fort Collins, Colorado.

Software dictionary

Software	Description
ESRI ArcGIS Desktop	ModelBuilder: outline analysis process and track changes in analysis methods. Cartographic tools: mapping and visualization.
ArcGIS Spatial Analyst extension	Map algebra for the overlay summation; tabulate area for the data summaries by watershed; zonal statistics for calculating mean score by watersheds.
ArcGlobe 9.2	3D visualization of overlay process.

Additional resources

Resource	Description and source
EPA Source Water Protection Program	Computed the drinking water consumers by 8-digit HUC using the EPA's Public Drinking Water System (PWS) data.
Martina Barnes, Regional Planner, USDA Forest Service, Northeastern Area State and Private Forestry	Project manager and technical support.
Albert H. Todd, watershed program leader (now assistant director, Ecosystem Services and Markets), USDA Forest Service, Northeastern Area State and Private Forestry	Technical support.
Paul K. Barten, PhD, Associate Professor, University of Massachusetts, Amherst	Designed the forests, water, and people analysis; edited water consumer data for the large river watersheds.

Recipe for map-building success

To quantify the forest-to-faucet connection, forest ecologists developed four indices to achieve the final map product using nine key input data layers. The assessment displays maps for 540 8-digit hydrologic unit code (HUC) watersheds in the northeastern United States using a four-step process to describe current and future conditions.

Step 1: Create the "ability to produce clean water index" (APCW) layer

Step 1 is an evaluation of physical and biological factors that serves as an index of the ability to produce clean water index for each watershed. This index provides a comparative ranking of predicted water quality and watershed integrity. Six 30-meter raster data layers form the basis for this analysis, including the percent of forest land, percent of agricultural land, percent of riparian forest cover, soil erodibility, road density, and housing density.

Each of these six data layers is rated through a query process generated by a GIS analyst from one to four with one being low and four being very high. To summarize by watershed, all layers were combined with the GIS to create the new APCW raster layer having a single score with a maximum of twenty-four and a minimum score of six.

Step 2: Create the "importance of watersheds for drinking water supply" layer

Step 2 shows the total water consumers served by surface water in each watershed and ranks these watersheds in terms of their importance in providing drinking water to the greatest number of people.

The "importance of watersheds for drinking water supply" layer was created by combining the results of the APCW layer created in step 1, with surface water use data from the U.S. Environmental Protection Agency's (EPA) Safe Drinking Water Information System.

  a. The total drinking water consumption was summed for each watershed and then divided by the watershed area. This result was then divided into ten quantiles, with the first quantile receiving a score of 10 and the tenth quantile receiving a score of 1.
  b. The APCW layer created in step 1 was also divided into ten quantiles, with the first quantile receiving a score of 10 and the tenth quantile receiving a score of 1.
  c. Both these layers were combined to create the "importance of watersheds for drinking water supply" layer to yield a score ranging from 2 to 20.

Step 3: Create the "importance of watersheds and private forests" layer

Step 3 highlights watersheds that are both critical for water supply to large population and contain a high percentage of unprotected private forest lands.

The "importance of watersheds and private forests" layer combines the results of step 2 with the watershed's percent of private forest cover to highlight those areas important for surface water drinking supply that contain private forest lands. Only permanently protected lands (federal, state, county, local, or permanent conservation easements) were considered protected and all other lands were considered unprotected. The percent private forest by watershed was divided into ten quantiles, and then combined with the results of step 2 to yield a total score ranging from 3 to 30.

Step 4: Create the "forest conversion pressure" layer

Step 4 highlights areas where the greatest development pressure threatens private forests that are important to the protection of surface drinking water supplies.

The "forest conversion pressure" layer combines the results of step 3 with the development pressure of future housing density change on forested areas. Development pressure was calculated by subtracting housing density in the year 2000 from the predicted housing density in 2030. If housing density increases from rural to exurban, rural to suburban/urban, or exurban to suburban/urban between 2000 and 2030, development pressure was said to occur. The total acreage of land under development pressure in the watershed was divided by watershed area, divided into ten quantiles, and then combined with the results of step 3 to yield a total score ranging from 4 to 40.

The final dataset was mapped using a blue to red gradation, where blue represented the most important surface drinking water watersheds at risk for future forest conversation.

Conclusion

This map shows the results of the interaction among forests, water, and people. By looking at these relationships on a landscape level, priorities for forest, water, and land management can be better determined. The unique results of this analysis can guide strategies for forest land protection, outreach, and technical assistance to municipal water providers.

Improving sustainability planning in Brazilian eucalyptus forests

19 CHAPTER

Richard Mendes Dalaqua

Suzano Papel e Celulose (also known as Suzano Pulp and Paper) is the largest integrated manufacturer of eucalyptus pulp and paper in Latin America and the largest producer of Forest Stewardship Council (FSC)-certified eucalyptus pulp in the world. Suzano has a long history of producing paper from eucalyptus, beginning with the development of plantations in 1924. In the mid-1960s, it was the first company to use eucalyptus cellulose on an industrial scale.

Suzano owns 1.1 million acres of land in five Brazilian states, including São Paulo, Bahia, Espírito Santo, Minas Gerais, and Maranhão. Of this land, close to half a million acres are set aside solely for eucalyptus plantations.

The Brazilian forest code requires that areas with slopes greater than 45 degrees are to be protected from harvesting and native vegetation is to be kept intact. All landowners must classify areas with slopes above 45 degrees for permanent preservation. In a number of cases, however, Suzano had established plantations long before the law and regulations were created. As a result, the company must now restore plantations on slopes above 45 degrees with tropical rain forest vegetation. Additionally, all slopes greater than 45 degrees must be identified and removed from any future harvest planning activities.

This map illustrates one of Suzano's 6,700-acre properties in the eastern state of São Paulo. The map was generated to identify slopes greater than 45 degrees and to show which of these areas needs to be restored back to native vegetation.

GIS specialists used aerial photographs at a scale of 1:300,000 to create contours, which were then transferred to the GIS to create a digital elevation model. They created a triangulated irregular network, or TIN, for the slope analysis of the property. Using the slope function, analysts calculated the maximum rate of change between contours representing elevation. To improve the visual quality of the map, the areas above 45 degrees were superimposed on a relief map to give viewers a 3D perspective.

A visual solution

Prior to the establishment of national regulations that protect forests on slopes above 45 degrees from development, Suzano converted a number of areas from native vegetation to eucalyptus plantations. As the code was implemented, foresters could not easily demarcate slope values over a large land base until GIS technologies became available to commercial forestry companies.

After investing in geospatial technologies, Suzano used GIS to create this map to delineate affected areas. This approach is critical for the field operations team to visualize then calculate the total acres of applicable slopes and plan and budget for converting eucalyptus plantations back to natural tropical rainforest vegetation.

This work helped the company in their efforts to meet the cellulose requirements for paper production. GIS helped Suzano quantify the size of new commercial land needed to be purchased as a replacement.

This map presents the different uses of soil, hydrography, elevation, and slope above 45 degrees for Suzano's property in the eastern state of São Paulo. With this information, it was possible to create a slope class map highlighting 45 degrees or more of inclination for the property.

USO DO SOLO E DECLIVIDADE
MAIOR QUE 45° FAZENDA MONTES CLAROS

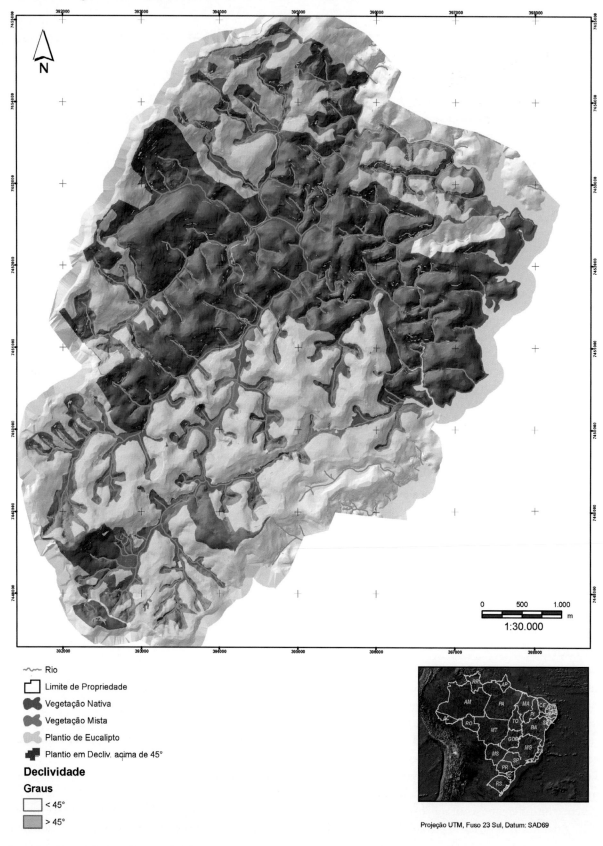

Data courtesy of Suzano Papel e Celulose.

Resource tables

Data dictionary

General data description	Data sources
Digital elevation model	Restitution of aerial photogrammetry.
Eucalyptus plantation	Suzano Papel e Celulose.
Hydrography	Suzano Papel e Celulose.
Native vegetation	Suzano Papel e Celulose.
Property boundary	Suzano Papel e Celulose.

Software dictionary

Software	Description
ESRI ArcGIS Desktop	Data management, spatial analysis, cartographic production.
ArcGIS 3D Analyst extension	TIN, slope, hillshade analysis.
ArcGIS Spatial Analyst extension	Raster analysis including reclassify, convert raster to vector.
SQL Server 2000	Relational database management system.

Additional resources

Resource	Description and source
ModelBuilder	Automated polygon generation of permanent areas of preservation.
GIS manager and cartographer	Able to use ArcGIS 3D Analyst and Spatial Analyst extensions.

Recipe for map-building success

Step 1: Define study area

The study area was selected based upon the variability of relief and the likelihood of identifying slopes above 45 degrees. The property was delineated according to legal boundaries of ownership.

Step 2: Organize data

Data necessary for analysis was obtained and loaded into a server database with ArcSDE technology. This data includes contours, hydrography, and forest inventories.

Step 3: Create TIN

A triangulated irregular network (TIN) was created to represent the topographic surface in 3D using Suzano's contour dataset as the elevation source. TINs are a type of vector-based geographic data created by triangulating points and connecting them with edges to form a network of triangles. Using a TIN, the mapmaker can show topographic surface at varying levels of resolution while efficiently storing data.

Rivers, streams, and roads were also used to capture abrupt changes in topography. This technique, called hard breaklines, represents a discontinuity in the slope of the terrain.

Step 4: Create slope layer

The TIN was used as an input to create a new slope layer. The generation of a slope feature class creates an output polygon classified according to the angle of inclination between the surface and horizontal plan attributes stored in the TIN. The angle used to classify slope is the maximum rate of elevation change in each triangle. Suzano GIS specialists created the output feature class using degrees; however, percent slope can also be used.

Step 5: Determine plantation areas above 45 degrees

With ArcMap, GIS analysts queried the slope feature class to select and display only those polygons over 45 degrees. Using the intersect tool, the geometric intersection between the input features slope and eucalyptus plantations derived from the forest inventory map are analyzed. Features that overlap in both feature classes are then output to a new polygon feature that identifies all the slopes greater than 45 degrees and are within the boundaries of a eucalyptus plantation.

The area in hectares for each plantation and for the total area affected in the property is now available by calculating the total area and the location easily identified on the screen using ArcMap.

Step 6: Create final map

ArcMap was used to create the final cartographic product to be used by Suzano forest planners and also shared with regulatory agencies.

The GIS specialist organized features, described the legend, and selected symbologies, including color palettes, line types, and line thicknesses to create a visually appealing map product that also looks real. To add a 3D element to the display, a hillshade was rendered with 60 percent transparency. Streams, rivers, roads, and eucalyptus plantations were added to provide additional and meaningful content.

After the map was created, it was printed as a PDF file and sent to forest and environmental operations managers. Hard-copy plots were also produced for field use.

Conclusion

Prior to the availability of GIS, terrain slopes were presented using contours and color shading to indicate slope. This made it difficult to quantify specific slope values and required considerable interpretation.

The use of GIS, however, has provided a significant enhancement to manual methods both in terms of accurate quantification of slope and location, but also in improved visualization. The inclusion of hillshade with transparency offers clarity to defining and displaying complex data, providing for easy identification of classified areas in a visually pleasing manner.

This type of analysis facilitates both strategic planning activities as well as tactical work conducted in the field. For strategic planning, the entire property is analyzed and attributed according to whether an area needs to be protected or rehabilitated into natural vegetation. By determining the location of these areas and calculating the size of each area classified as protected, Suzano managers can determine the cost required to restore these areas as prescribed under Brazilian law.

Author information

Chapter 1
Carlos Souza Jr., PhD, senior researcher, Imazon
Belém-Pará, Domingos Marreiros Street, 2020, Brazil, souzajr@imazon.org.br, 55-91-3182-4000

Amintas Brandão Jr., assistant researcher, Imazon
Belém-Pará, Domingos Marreiros Street, 2020, Brazil, brandaojr@imazon.org.br, 55-91-3182-4000

Marco Lentini, MSc, associate researcher, Imazon
Belém-Pará, Domingos Marreiros Street, 2020, Brazil, lentini@imazon.org.br, 55-91-3182-4000

Chapter 2
Charles H. (Hobie) Perry, PhD, research scientist, USDA Forest Service, Northern Research Station
1992 Folwell Avenue, St. Paul, MN 55108, charleshperry@fs.fed.us, 651-649-5191

Mark D. Nelson, PhD, research forester, USDA Forest Service, Northern Research Station
1992 Folwell Avenue, St. Paul, MN 55108, mdnelson@fs.fed.us, 651-649-5104

Ronald J. Piva, forester, USDA Forest Service, Northern Research Station
1992 Folwell Avenue, St. Paul, MN 55108, rpiva@fs.fed.us, 651-649-5150

Chapter 3
Perttu Anttila, PhD, researcher, Finnish Forest Research Institute (Metla)
P.O. Box 68, FI-80101 Joensuu, perttu.anttila@metla.fi, 358-50-391-3088

Lauri Sikanen, PhD, professor, Faculty of Forest Sciences, University of Joensuu
P.O. Box 111, FI-80101 Joensuu, lauri.sikanen@joensuu.fi, 358-13-251-3636

Dominik Röser, researcher, Finnish Forest Research Institute (Metla)
P.O. Box 68, FI-80101 Joensuu, dominik.roser@metla.fi, 358-50-391-3266

Juha Laitila, researcher, Finnish Forest Research Institute (Metla)
P.O. Box 68, FI-80101 Joensuu, juha.laitila@metla.fi, 358-10-211-3255

Author information

(Chapter 3 continued)
Timo Tahvanainen, development manager, Joensuu Science Park Ltd.
Länsikatu 15, Joensuu, FI-80110, timo.tahvanainen@carelian.fi

Heikki Parikka, researcher, Finnish Forest Research Institute (Metla)
P.O. Box 68, FI-80101 Joensuu, heikki.parikka@metla.fi, 358-10-211-3051

Chapter 4
Nate Anderson, PhD, postdoctoral research forester, USDA Forest Service, Rocky Mountain Research Station
800 East Beckwith Avenue, Missoula, MT 59801, nmanderson@fs.fed.us, 406-542-4158

Eddie Bevilacqua, PhD, associate professor, SUNY College of Environmental Science and Forestry, Department of Forest and Natural Resources Management, 301 Bray Hall, 1 Forestry Drive, Syracuse, NY 13210, ebevilacqua@esf.edu, 315-470-669

René Germain, PhD, associate professor, SUNY College of Environmental Science and Forestry, Department of Forest and Natural Resources Management, 316 Bray Hall, 1 Forestry Drive, Syracuse, NY 13210, rhgermai@esf.edu, 315-470-6698

Chapter 5
Andriy V. Zhalnin, PhD, forestry GIS specialist, Department of Forestry and Natural Resources, Purdue University
715 West State Street, West Lafayette, IN 47907, lesovod@yahoo.com, 765-496-3263

Richard L. Farnsworth, PhD, associate professor, GIS-Based Decision Support, Department of Forestry and Natural Resources
Purdue University, 715 West State Street, West Lafayette, IN 47907, rlfarnsw@purdue.edu, 765-496-3245

Shorna Broussard Allred, PhD, associate professor, Department of Natural Resources, Cornell University
122C Fernow Hall, Ithaca, NY 14853, srb237@cornell.edu, 607-255-2149, www.human-dimensions.org

Brett A. Martin, GIS coordinator, Indiana Department of Natural Resources, Division of Forestry
402 W. Washington Street, Room W296, Indianapolis, IN 46204, bmartin@dnr.in.gov, 317-232-4106

Chapter 6
Greg C. Liknes, MS, research physical scientist, USDA Forest Service, Northern Research Station
1992 Folwell Avenue, St. Paul, MN 55108, gliknes@fs.fed.us, 651-649-5192

Mark D. Nelson, PhD, research forester, USDA Forest Service, Northern Research Station
1992 Folwell Avenue, St. Paul, MN 55108, mdnelson@fs.fed.us, 651-649-5104

Brett J. Butler, research forester, Family Forest Research Center, USDA Forest Service, Northern Research Station and Family Forest Research Center, 160 Holdsworth Way, Amherst, MA 01003, bbutler01@fs.fed.us, 413-545-1387

Chapter 7
Paul D. Graham, urban forester/horticulturalist/GIS specialist, City of Florence, Department of Urban Forestry and Horticulture
2806 Darby Drive, Florence, AL 35630, pgraham@florenceal.org, 256-718-5025

Christopher J. Holder, intern GIS specialist, Hamilton County, GIS Department
1250 Market Street, Chattanooga, TN 37402, christopherh@hamiltonTN.gov, 423-209-7760

Chapter 8

Marius Dumitru, GIS specialist, Forest Research and Management Institute (ICAS)
Sos. Stefanesti 128, 077190 Voluntari, Judetal, Ilfov, Romania, mariusd@icas.ro, +40740277880

Marius Daniel Nitu, GIS specialist, Forest Research and Management Institute (ICAS)
Sos. Stefanesti 128, 077190 Voluntari, Judetal, Ilfov, Romania, dannitu@icas.ro, +40745906570

Gheorghe Marin, forest engineer, Forest Research and Management Institute (ICAS)
Sos. Stefanesti 128, 077190 Voluntari, Judetal, Ilfov, Romania, ghmarin@icas.ro, +40744650982

Chapter 9

Heinrich Goetz, PhD candidate, Institute of Applied Science, University of North Texas
2025 Crestwood Place, Denton, TX 76209, heinrichgoetz@my.unt.edu, 940-594-5657

Chapter 10

Dana Kao, PhD candidate, Laboratory of Forest Management, Department of Forests and Forest Product Science
Graduate School of Bioresource and Bioenvironmental Sciences, Kyushu University
6-10-1, Hazozaki, Higashiku, Fukuoka, 812-8581, Japan, kaodana@gmail.com, 855-12-540-009

Tsuyoshi Kajisa, PhD, assistant professor, Faculty of Agriculture, Kyushu University
6-10-1, Hazozaki, Higashiku, Fukuoka, 812-8581, Japan, kajisa@ffp.kyushu-u.ac.jp, 81- 92-642-2869

Nobuya Mizoue, PhD, associate professor, Faculty of Agriculture, Kyushu University
6-10-1, Hazozaki, Higashiku, Fukuoka, 812-8581, Japan, mizoue@ffp.kyushu-u.ac.jp, 81-92-642-2866

Shigejiro Yoshida, PhD, professor, Faculty of Agriculture, Kyushu University
6-10-1, Hazozaki, Higashiku, Fukuoka, 812-8581, Japan, syoshida@agr.kyushu-u.ac.jp, 81- 92-642-2865

Chapter 11

Matthew P. Peters, GIS technician, USDA Forest Service, Northern Research Station
359 Main Road, Delaware, OH 43015, mpeters@fs.fed.us, 740-368-0145

Louis R. Iverson, PhD, landscape ecologist, USDA Forest Service, Northern Research Station
359 Main Road, Delaware, OH 43015, liverson@fs.fed.us, 740-368-0097

Anantha M. Prasad, MS, ecologist, USDA Forest Service, Northern Research Station
359 Main Road, Delaware, OH 43015, aprasad@fs.fed.us, 740-368-0103

Chapter 12

David Yun, lecturer, Natural Resources Management Department, California Polytechnic State University
1 Grand Ave, San Luis Obispo, CA 93407, dyun@calpoly.edu, 805-781-7189

Chapter 13

Holli Howard, director of geographic resources, Casey Trees
1123 11th Street, NW, Washington, D.C. 20001, hhoward@caseytrees.org, 202-349-1905

Mike Alonzo, GIS specialist, Casey Trees
1123 11th Street, NW, Washington, D.C. 20001, malonzo@caseytrees.org, 202-349-1897

Chapter 14

Bruce Blackwell, MSc, RPBio, RPF, BA Blackwell and Associates Ltd.
3087 Hoskins Road, North Vancouver, British Columbia, V7J 3B5, bablackwell@bablackwell.com, 604-985-8769

Franz Feigl, MSc, senior GIS analyst, Forest Ecosystem Solutions Ltd.
227-998 Harbourside Drive, North Vancouver, British Columbia, V7P 3T2, ffeigl@forestecosystem.ca, 604-998-2222

Chapter 15

Weimin Xi, PhD, research scientist, Knowledge Engineering Laboratory, Department of Entomology, Texas A&M University
College Station, TX, xi@tamu.edu, 979-845-9736

Lei Wang, PhD, assistant professor, Department of Geography and Anthropology, Louisiana State University
Baton Rouge, LA, leiwang@lsu.edu, 225-578-8876

Maria D. Tchakerian, PhD, assistant research scientist, Knowledge Engineering Laboratory, Department of Entomology
Texas A&M University College Station, TX, mtchakerian@tamu.edu, 979-845-9735

Robert N. Coulson, PhD, professor and director, Knowledge Engineering Laboratory, Department of Entomology and Department of Ecosystem Science and Management, Texas A&M University, College Station, TX, r-coulson@tamu.edu, 979-845-9725

Chapter 16

Clark Labs, Clark University
950 Main St., Worcester, MA, www.clarklabs.org, 508-793-7526

Chapter 17

Chris Zanger, MS, fire research analyst, The Nature Conservancy, Central Oregon Office
115 NW Oregon Avenue, Bend, OR 97701, 541-388-3020

Amy Waltz, PhD, fire ecologist, The Nature Conservancy, Central Oregon Office
115 NW Oregon Avenue, Bend, OR 97701, 541-388-3020

Chapter 18

Rebecca Lilja, GIS specialist, USDA Forest Service, Northeastern Area State and Private Forestry
271 Mast Road, Durham, NH 03824, 603-868-7627, rlilja@fs.fed.us

Chapter 19

Richard Mendes Dalaqua, geoprocessing specialist, Suzano Papel e Celulose, Rua. Dr. Prudente de moraes
4006, Bairro Areião, Suzano, São Paulo, 08613-900, rdalaqua@suzano.com.br, 55-11-3636-5877